D1797075

Genomics and Public Health

Genomics and Public Health

Legal and Socio-Ethical Perspectives

Edited by

Bartha Maria Knoppers

MARTINUS NIJHOFF PUBLISHERS
LEIDEN / BOSTON

A Cataloging-in-Publication record for this book is available from the Library of Congress.

Printed on acid-free paper.

ISBN 10 90 04 15559 7
ISBN 13 978 90 04 15559 6
© 2007 by Koninklijke Brill NV, Leiden, The Netherlands

Koninklijke Brill NV incorporates the imprints Brill, Hotei Publishers,
IDC Publishers, Martinus Nijhoff Publishers and VSP.

www.brill.nl

Printed and bound in The Netherlands.

Acknowledgements

The 4th International DNA Sampling Conference on: *"Genomics & Public Health"* in June 2006 was a memorable event and the foundation for this book. For this accomplishment, we wish to thank Marie-Hélène Régnier and Matthieu Layes in particular, and the Genetics & Society team led by Denise Avard – an excellent effort as always!

The book owes its quality not only to the contributions of the authors, but also to the superb editing of Alexandra Saginur and the professionalism of Emmanuelle Lévesque and Cecile Dubeau.

Merci à tous!

Bartha Maria Knoppers, Chair
4th International DNA Sampling Conference
Canada Research Chair in Law & Medicine
Centre de recherche en droit public (CRDP)
Université de Montréal

A SPECIAL THANKS TO OUR SPONSORS:

Genome Quebec & Genome Canada

AND:

IIREB – International Institute of Research in Ethics and Biomedicine
CDC – Centers for Disease Control and Prevention (Co-Organizer)
Canada Research Chair in Law and Medicine
HUMGEN, www.humgen.umontreal.ca
Ministère de la Santé et des Services Sociaux, Québec
Health Law Institute, University of Alberta
CGKP – Cambridge Genetics Knowledge Park (Co-Organizer)
RMGA – Réseau de Médecine Génétique Appliquée de FRSQ, Québec
CIHR – Canadian Institutes of Health Research – Institute of Genetics
CIHR – Canadian Institutes of Health Research, Institute of Health Services and Policy Research
University of Toronto, Joint Centre for Bioethics
Public Health Agency of Canada
Department of Justice, Canada
Wellcome Trust, United Kingdom

Table of Contents

PART III
GENOMICS AND PUBLIC HEALTH: CURRENT APPROACHES AND FUTURE PERSPECTIVES

PART I

POPULATION SCREENING: ISSUES, REALITIES AND POSSIBILITIES

A – Introduction:
Expansion of Screening?

J. Gerard LOEBER

Dutch National Institute for Public Health, The Netherlands

All parents want their babies to be and remain in good health. They want—no, they expect—to be informed by the healthcare system about anything that conflicts with this expectation. Over the last 40 years, neonatal screening technology has evolved tremendously, and it is still making giant leaps. In the beginning, there were just simple tests for diseases like amino acidopathies, phenylketonuria being the classic example. The fact that the PKU phenotype may be the result of a variety of genotypes was discovered years later.

The success of the Guthrie bacterial inhibition assay using filter paper blood samples was the key to the expansion of screening programmes. At first, there was the addition of one other disease, congenital hypothyroidism (CH), which was depicted as being as simple as PKU but in fact has been shown to have various ethiologies. PKU and CH were and are still important because of their impact on lifelong suffering, if undetected, and their relatively easy means of treatment. For these reasons, virtually all neonatal screening programmes have started with those two diseases, and many still are limited to them. Advances in biochemical techniques in the 1970s and 1980s made it possible to detect many more diseases in the same blood spot material. Many of them are relatively easy to treat and simple to explain to parents and professionals.

The development of tandem mass spectrometry and methods for genetic mutation analysis in the 1990s have complicated things. Not only did the number of diseases rise exponentially, but also the concept of carrier detection made life difficult for the screening professionals. How to explain all possible variations and chances as concerns the development of rare diseases to parents who just want to enjoy the arrival of a seemingly healthy child? And then, not only for this particular heterozygote child, but also for future children who might or might not be homozygous for some disorder?

Now we are on the brink of an even more complicated, yet interesting era. Not only can we determine with some probability that a child autonomously will develop a certain disease, but we can also predict the susceptibility for such development, which in many cases may be prevented by changing lifestyle, living environment, or diet. We have become part of the knowledge paradox: the more we know, the more we know we do not know. Through having gained and as we still gaining this knowledge, we must ask ourselves how to deal with it. How to counsel the parents, to ask their consent? How to educate the professionals for them to do this job properly? How much should ethics be involved in steering such a programme? Is our community ready for these questions? We might come to the conclusion that we have opened up Pandora's Box!

Expansion of Newborn Screening: Current Achievements and New Prospects

Bridget WILCKEN

The New South Wales Newborn Screening Programme, The Children's Hospital at Westmead and the University of Sydney, Australia.

1. THE EARLY HISTORY OF NEWBORN SCREENING

In considering the present and future of newborn screening, it is useful to remind ourselves of its beginnings. Newborn screening could probably be said to have started in 1961, with a letter to the editor of *The Journal of the American Medical Association* from Dr. Bob Guthrie describing his screening test for phenylketonuria (PKU), a few months before his seminal paper in the *Journal of Pediatrics.*[1] By 1963, the state of Massachusetts had passed a law mandating this test; they were closely followed by Oregon and other states. While there was some opposition, the test was mandatory in most states by 1966. The era of newborn screening had begun. The original "Guthrie" test was a bacterial inhibition assay, and soon other similar individual tests were introduced for a few very rare disorders, plus an enzyme assay for galactosaemia. Some 12 years later, another major step forward was a test for congenital hypothyroidism.[2] During the 1970s, other disorders were considered but not widely tested for.[3] In the 1980s, blood-spot tests for cystic fibrosis were taken up enthusiastically by a few screening programmes, despite a lot of opposition,[4] as were tests for congenital adrenal hyperplasia, biotinidase deficiency, haemoglobinopathies, and neuroblastoma. Not all of these screening programmes were to last.

The 1980s saw the institution of the first well-designed trials, of cystic fibrosis[5] and of neuroblastoma.[6] There is now clear scientific evidence of efficacy of screening for the former [CDC], but equally valuably, infant screening for neuroblastoma was shown unequivocally to be ineffective. It is worth noting that the trials of CF screening were the only double blind, randomised trials (pseudo-randomised in the case of the Wales West-Midland UK trial[7]) to be carried out to investigate newborn screening. Despite the trials, over those three decades, newborn screening had been in the main a quiet, somewhat back-water, exercise.

Two things were to change this. Firstly, the demonstration that DNA could easily be extracted from dried blood spots and used in screening: this was initially applied to cystic fibrosis and haemoglobinopathies.[8] The second major advance was the introduction of electrospray ionisation tandem mass spectrometry (MSMS),[9] enabling multiplex testing for many different disorders in one test. This was termed "expanded newborn screening."[10] Newborn screening had suddenly become exciting, and there was a surge of people and institutions wishing to become involved. Also, at least in the United States, there was much public pressure exerted for the rapid adoption of extended newborn screening.

2.   EXPANDED NEWBORN SCREENING

To assess how much has been achieved by this expansion of newborn screening, we first need to revisit the overarching aim of newborn screening. Put simply, the primary aim is to detect babies with medically significant, treatable disorders for whom presymptomatic diagnosis and treatment would be beneficial, and to facilitate that treatment. There may also be more controversial secondary aims, such as a benefit to the family to enable subsequent reproductive choices. To realise these aims, a number of processes need to be in place. These of course include appropriate laboratory services and follow-up procedures. In addition, there must be methods for the sensible selection of disorders to be included in a screening panel, as well as case definitions; development of educational material for professionals and the public; evaluation of outcomes; strategies for harm minimisation; and promotion of research. The latter is doubly important as newborn screening is expanding in this dramatic and relatively sudden way.

Expanded newborn screening by MSMS has been used now for over a decade. Developed largely in North Carolina, the first screening programmes were in a private facility (NeoGen Screening, now called Pediatrix Analytical) in Pennsylvania, and the first state-wide programme in the public arena was undertaken in North Carolina in 1997,[11] although the first continuous state-wide programme run "in house" was the programme in New South Wales, Australia, which started in 1998.[12] Now, in 2006, tandem mass spectrometry screening has been widely undertaken. It is universal in Australia, and in several European countries, including Germany, Austria, and Belgium, and in over 40 of the 50 states in the USA. It has also been adopted in some countries in the Middle East, and Far East, often with partial coverage.

3.   CURRENT ACHIEVEMENTS

3.1   *High Sensitivity and Specificity*

At present, MSMS newborn screening is used almost exclusively for the detection of selected disorders of amino acid and organic acid metabolism, and fatty acid oxidation defects. Early papers indicated the likely scope of such screening[13] and now several screening programmes have reported comprehensively on their results,[14] two documenting carefully all the known missed cases. These reports indicate that secure early diagnosis—close to 100% sensitivity—is achievable for a range of disorders with a low overall false positive rate. Additionally, there is certainty that some disorders have far short of 100% sensitivity, and in the middle are disorders whose rarity has made the sensitivity not yet clear [table 1]. Complicating this picture is the lack of agreement on case-definitions (see below) and unclear ascertainment of missed cases. However, detection rates have been very satisfactory overall and have been achieved with very low false positive rates. In New South Wales, the overall false positive rate for MSMS, including PKU, is 0.2%, and other programmes have false positive rates of 0.33% or less.[15]

TABLE 1: RELIABILITY OF DETECTION OF DISORDERS BY CURRENT TANDEM MASS SPECTROMETRY NEWBORN SCREENING, ASSUMING A LOW OVERALL FALSE POSITIVE RATE OF UP TO 0.33%.

RELIABILITY OF DETECTION APPARENTLY HIGH	RELIABILITY OF DETECTION UNCLEAR (MOST APPARENTLY HIGH)	DETECTION UNRELIABLE
Phenylketonuria	Argininosuccinic aciduria	Tyrosinaemia type I (except with SA assay)
MSUD neonatal onset	GA I	Homocystinuria (CBS) B6-responsive
Citrullinaemia type I	Homocystinuria (CBS) B-6 non-responsive	MSUD intermediate/intermittent
Neonatal onset organic acidaemias	3-MCCC deficiency	Non-ketotic hyperglycinaemia
MCAD deficiency	Late-onset organic acidurias	Beta-ketothiolase deficiency
	VLCAD deficiency	Cobalamin C defect
	LCHAD deficiency	Carnitine uptake defect
	TFP deficiency	

MSUD – maple syrup urine disease; MCAD – medium-chain acyl-CoA dehdrogenase; GA I – glutaryl CoA dehdrogenase deficiency; CBS – cystathionine beta-synthase; 3-MCCC – 3-methylcrotonyl CoA carboxylase; VLCAD – very-long-chain acyl-CoA dehydrogenase; LCHAD – long-chain 3-hydroxyacyl-CoA dehydrogenase; TFP – trifunctional protein deficiency; SA – succinylacetone

It is important for screening programmes to take cognizance of what they can and cannot hope to achieve. Reports of missed cases and reports of disorders which seem not to be able to be detected with 100% certainty are thus very important. For example, we have recently reported the inability to detect cases of intermittent maple syrup urine disease[16] and non-ketotic hyperglycinaemia. As another example, there is general agreement among screeners that in using the present approach, it is not possible to detect patients with pyridoxine-responsive homocystinuria (cystathionine β-synthase deficiency), although patients with the more severe pyridoxine-non-responsive forms can be detected.[17] For some people, this poses a problem: should one include for screening a disorder for which the sensitivity is low—say 50%, as in the case of homocystinuria? Because with MSMS, screening additional disorders can be included with virtually no increase in overall cost, so it does seem unreasonable to deny an advantage to those babies who could be detected and benefit from early detection because some cannot be so detected. The problem may be a risk-management one in a litigious society, but the ethical position seems clear.

3.2   Organisation of Follow-up and Treatment Options

Many newborn screening laboratories are operating as part of a public health laboratory, with little exposure to the biochemical and clinical services which take care of confirmatory testing and treatment. In the past, with screening for a limited number of well-understood conditions, this did not seem to have adverse consequences. With the sudden expansion of newborn screening to include rare, less-well understood metabolic disorders, protocols have been developed to recommend which disorders should be included in a screening programme, and educational material has been developed for the professionals to guide follow-up testing and treatment. Not all of this has had a smooth course. Australia has been lucky to have newborn screening already well integrated with biochemical genetics and clinical services, and this seems to be an ideal situation. All Australian screening programmes are located within teaching hospitals, and co-located with the biochemical and clinical services. Elsewhere, the "forced marriage" of these services may have been initially difficult, but is likely to have very positive consequences: there will be much more awareness, bilaterally, of screening programmes and clinical services and their needs and requirements. This "marriage" is also likely to improve diagnostic and clinical services overall, as they will operate less in an isolated situation; this could be a trigger for the start of specialised clinics, as occurred some time ago in Normandy, with cystic fibrosis screening (G. Travert, personal communication). And certainly newborn screening services as a whole will be improved as they face up to more complex tasks with well-worked-out processes.

4.   PROBLEMS STILL REMAIN

While there are undoubted current and potential achievements, remaining problems need to be tackled. There is not overall agreement about which disorders should be included in the screening suite; there may be problems of over-diagnosis; follow-up and treatment is not yet always appropriate; and there are great difficulties in evaluation. These issues will be discussed in turn.

4.1   *Choosing Which Disorders to Include*

There have been several sets of criteria chosen to validate inclusion of a disorder in a screening programme. The most quoted is the beautifully written document of Wilson and Jungner.[18] This was published in 1968 by the World Health Organization, before any blood spot newborn screening except for PKU was contemplated. It was in the main related to general health screening and screening for infectious diseases and cancers. The 10 principals are largely but not completely relevant to newborn screening. Since then, other sets of criteria have been proposed: for example, the United Kingdom Guidelines.[19] But put simply, to include a new disorder in a newborn screening programme, there needs to be a demonstrated *likely* benefit (there is usually no unequivocally demonstrated benefit before extensive pilot programmes have been undertaken), and the costs and harms of all kinds should be considered and found to be reasonably balanced against the benefits. There is a fear that once a screening programme has been started, nothing will stop it. Certainly programmes do exist which have remained un-validated for years. But programmes have also been stopped in the past when benefit could not be demonstrated, as exemplified by neuroblastoma screening,[20] urine screening[21] and others. With regard to MSMS newborn screening, different jurisdictions have chosen different methods of selecting disorders and come up with entirely different solutions.[22]

The United States model was to invite various professionals—laboratory directors, clinicians, administrators and so forth—to fill in a complex evaluation, with a points system for burden of disease, performance of the screening test, availability of diagnostic testing and treatment centres or specialists, and likely outcome. The model appeared very flawed: the assignment of points seemed arbitrary, and several of the disorders included for evaluation were so rare (for one disorder, dienoyl CoA-reductase deficiency, there had only been one published case) that many respondents could not have had any experience that could relate to most of the questions. However, a literature review by one or two specialists was included for each disorder, and the overall result has been very helpful.[23] The USA recommended the inclusion of 29 core disorders to be screened for, 20 detectable by MSMS screening, and a further 25 "secondary targets" (22 by MSMS)—disorders likely to be found by screening for the core disorders recommended, and which should be reported—so an overall list of 54 disorders. Not all states have yet (in mid-2006) implemented these recommendations.

A German model was a consideration by a committee of which disorders would be covered by universal insurance if detected by newborn screening. This resulted in a panel of 14 metabolic and endocrine disorders that should be included in screening programmes. Laboratories were instructed that results which could indicate other disorders should not be reported. Indeed a translation of the document indicates that "The analysis of additional... disorders is not part of newborn screening. If the quantitation of analytes not necessary for the detection of the disorders listed can not be prevented due to technical reasons they have to be destroyed immediately. They must not be used, stored or transmitted to anybody."[24] The guidelines were to be reviewed two years after implementation. However, the ethical issues they pose to screeners are clear.

The United Kingdom has taken a rigorous approach. In the 1990s, two technical assessments were commissioned.[25] These came up with somewhat different conclusions, but on the whole, they supported expansion of newborn screening, citing medium-chain dehydrogenase (MCAD) deficiency and glutaric aciduria type I as being disorders in which early diagnosis was especially likely to provide benefit, with averted mortality and morbidity. Subsequently, a trial of screening for MCAD deficiency was started in 2004, with approximately half of the country screened and half not screened. Tandem mass spectrometry has not been used to screen for any other disorder, apart from PKU. This trial is due to report in 2008.

In Australia so far, the approach has been more *ad hoc*, with the individual programmes, based within state jurisdictions, making decisions separately. However, an overarching committee of the Human Genetics Society of Australasia and the Royal Australasian College of Physicians—a committee entirely composed of professionals involved in screening and clinical service—has met yearly for over 20 years to make recommendations. In Australia, we have been happy to include a number of conditions for which possible benefit is not clear-cut. This is mainly because of the demonstrated low false positive rate, and the integrated laboratory and clinical services which ensure rapid and accurate assessments of individual cases and minimize possible harm from parental distress and related factors. A research programme is under way to assess the outcomes.

4.2   *Over-diagnosis*

Over-diagnosis is common, indeed probably usual, in screening programmes. This is because, for many situations, it is impossible to draw a firm line between significant disease and mild variation from the norm. This is particularly true of genetic metabolic disorders, such as are diagnosed by MSMS, where mild gene variants may lead to slight biochemical derangement of no clinical significance. Over-diagnosis may occur for possibly serious conditions, but in the context of MSMS may also result in the discovery of cases of disorders most probably benign. With MCAD deficiency, a potentially fatal disorder of fatty acid oxidation, about twice as many cases are found by screening as ever present clinically.[26] Many but not all of these "extra" cases are genetically similar to those diagnosed after clinical presentation.[27] The cases missing from un-screened cohorts may have escaped any symptoms, due to epigenetic or environmental causes, or they may have been symptomatic, and even died, but remained undiagnosed. More concerning is the very frequent discovery of cases of disorders previously thought very rare. Two well documented examples are short-chain acyl-CoA dehydrogenase (SCAD) deficiency, and 3-methylcrotonyl CoA carboxylase (3-MCCC) deficiency. The symptomatology attributed to clinically detected SCAD deficiency has been heterogeneous,[28] and cases detected by the Australian screening programme have been entirely asymptomatic, without treatment. For 3-MCCC deficiency, numbers of affected mothers have been discovered because of transient metabolic abnormalities in their babies' screening test results.[29] All affected mothers and babies detected by screening in the programme in New South Wales have been asymptomatic, and for at least two large screening programmes, such cases have never previously been diagnosed clinically.[30] These problems are not fully resolved, but it seems likely that a number of subjects detected by newborn screening have been unnecessarily medicalised.

4.3   Follow-up

The expansion of newborn screening seems likely to improve prospects for good organization of confirmatory testing and follow-up for patients with these rare metabolic disorders, as implementation of expanded screening could well be a trigger for starting specialized metabolic clinics where none previously existed. While there is no evidence this would improve outcomes, it does seem likely, and merits evaluation. However, care must be exercised about the medicalisation problem referred to above. A fact sheet endorsed by the United States Maternal and Child Health Bureau about a rare disorder, isobutyryl CoA dehydrogenase deficiency, states that "little is known about the (clinical) effects. So far, symptoms have only been reported in one child." The sheet describes the symptoms as including cardiomyopathy, anaemia and poor growth. It goes on to state that "treatment is likely to be needed throughout life" and that this might include "medications" and a "low-valine food plan including medical foods."[31] This is another example of a disorder found mainly by newborn screening, with discovered patients asymptomatic, where there is a grave danger of over-treatment and the medicalisation of a benign condition. This would indeed be a grave harm.

4.4   *Measuring Clinical Effectiveness: Problems in Evaluation*

Such problems have recently been described.[32] Most importantly, the rarity of the disorders means that very large studies are needed. Finding more cases by screening than by clinical presentation could bias results in various ways, and might particularly suggest an advantage for outcomes if mildly affected not-at-risk patients are included in the screened and treated cohorts. One solution to this is to look at certain outcomes on a whole population basis. Completeness of ascertainment in unscreened and screened groups is also crucial if comparisons are to be made, as is comparability of treatment. Large follow-up and evaluation studies are under-way in Massachusetts,[33] Australia and North Carolina,[34] and there have been several more limited reports.[35]

5.   FUTURE PROSPECTS

5.1   *Lysosomal Storage Disorders*

The next newborn screening expansion seems likely to be for lysosomal storage disorders (LSDs). Novel treatments have been developed for several LSDs and other storage disorders in recent years.[36] Methods have been developed for identifying these disorders in dried blood spots[37] and for one disorder, Krabbe disease, (globoid-cell leukodystrophy), a screening programme may start soon. Results of pre-symptomatic treatment of Krabbe disease by umbilical cord blood transplant have been published,[38] and seem very promising. Certainly mortality and morbidity are, at the very least, greatly improved. It is not yet certain if, for this and other similar storage disorders such as Pompe disease (acid maltase deficiency), pre-symtomatic treatment might not be converting a devastating and rapidly fatal disorder into chronic, but still ultimately fatal disease. This needs to be taken into consideration. However for CF, ultimately life-shortening, screening has led to improved health, better quality of life, and possibly but not yet certainly an extended life-span,[39] and the same may be true for many LSDs.

5.2   *Adding New Disorders to a Screening Programme*

What could be added depends largely on four factors. Firstly, the discovery of new ways to detect disorders pre-symptomatically, either with new technology, or with new understanding of the pathophysiology leading to finding biochemical or molecular markers, would place disorders into a "screenable" category. Secondly, development of treatments might similarly affect a decision to screen, as in the instance of some LSDs referred to above. Thirdly, discovery of preventive measures would be an immensely powerful spur to screening for, for example, childhood cancers or type I diabetes. A fourth way in which disorders might be accepted for inclusion into a newborn screening programme would be a substantial shift in public opinion about what would be desirable to detect early. Here, the consideration might be newborn susceptibility testing (say, for diabetes) or screening for adult-onset diseases (eg for heart disease or, more controversially, for Huntington disease). Public opinion would be unlikely to support the latter. Susceptibility markers could be very important if preventive strategies could modify morbidity or mortality, but where lifestyles are affecting the onset of disease, there is as yet little evidence that behaviour modification is easy to achieve, and the prognostic reliability of susceptibility testing is uncertain. How would parents deal with weighing up the risks of cancer against the risks of vascular disease, psychiatric problems, accident-proneness, carrier status for monogenic disorders and so forth? Undoubtedly, microarrays for SNPs associated with various disorders, already available, will eventually become cheap enough to use in newborn screening, but uncertainties in genetic testing[40] are likely to remain.

6.   CONCLUSION

The immediate challenge for the newborn screening community is to be rigorous about the evaluation of existing programmes, a difficult but necessary task. For the future, it will probably not be a brave new world of neonatal screening for everything. Rather, it will be keeping up with medical advances and aiming to work on the efficient diagnosis of disorders that have new treatments (e.g., lysosomal storage disorders), working on developing a test where a new preventive measure is clear-cut, and carefully evaluating new screening possibilities for likely benefit and possible harm.

REFERENCES

1.   R. Guthrie & A. Susi, *A Simple Phenylalanine Method for Detecting Phenylketonuria in Large Populations of Newborn Infants*, 32 PEDIATRICS 338 (1963).
2.   J. H. Dussault et al., *Preliminary Report on a Mass Screening Program for Neonatal Hypothyroidism*, 86 J. PEDIATR. 670 (1975).
3.   Neonatal Screening for Inborn Errors of Metabolism (H. Guthrie et al. eds., 1980).
4.   B. Wilcken & G. Travert, *Neonatal Screening for Cystic Fibrosis: Present and Future,* 88 ACTA PAEDIATR. (supplement) 33 (1999).
5.   S. Chatfield et al., *Neonatal Screening for Cystic Fibrosis in Wales and the West Midlands: Clinical Assessment After Five Years of Screening*, 66 ARCH. DIS. CHILD. (special issue) 29 (1991); P. M. Farrell & E. H. Mischler, *Newborn Screening for Cystic Fibrosis. The Cystic Fibrosis Neonatal Screening Study Group,* 39 ADV. PEDIATR. 35 (1992).
6.   W. G. Woods et al., *Screening of Infants and Mortality Due to Neuroblastoma,* 346 N. ENGL. J. MED. 1041 (2002).
7.   Chatfield, *supra* note 5.
8.   E. R. McCabe et al., *DNA Microextraction from Dried Blood Spots on Filter Paper Blotters: Potential Applications to Newborn Screening*, 75 HUM. GENET. 213 (1987).
9.   D. H. Chace et al., *The Application of Tandem Mass Spectrometry to Neonatal Screening for Inherited Disorders of Intermediary Metabolism*, 3 ANNU. REV. GENOMICS HUM. GENET. 17 (2002); D. S. Millington et al., *Tandem Mass*

Spectrometry: A New Method for Acylcarnitine Profiling with Potential for Neonatal Screening for Inborn Errors of Metabolism, 13 J. INHERIT METAB. DIS. 321 (1990).

10. H. L. Levy, *Newborn Screening by Tandem Mass Spectrometry: A New Era,* 44 CLIN. CHEM. 2401 (1998).

11. D. M. Frazier et al., *The Tandem Mass Spectrometry Newborn Screening Experience in North Carolina: 1997-2005,* 29 J. INHERIT METAB. DIS. 76 (2006).

12. B. Wilcken et al., *Screening Newborns for Inborn Errors of Metabolism by Tandem Mass Spectrometry,* 348 N. ENGL. J. MED. 2304 (2003); V. Wiley et al., *Newborn Screening with Tandem Mass Spectrometry: 12 Months' Experience in NSW Australia,* 88 ACTA PAEDIATR. (supplement) 48 (1999).

13. D. H. Chace et al., *Rapid Diagnosis of Phenylketonuria by Quantitative Analysis for Phenylalanine and Tyrosine in Neonatal Blood Spots by Tandem Mass Spectrometry,* 39 CLIN. CHEM. 66 (1993); D. H. Chace et al., *Rapid Diagnosis of Maple Syrup Urine Disease in Blood Spots from Newborns by Tandem Mass Spectrometry,* 41 CLIN. CHEM. 62 (1995); D. H. Chace et al., *Rapid Diagnosis of Homocystinuria and Other Hypermethioninemias from Newborns' Blood Spots by Tandem Mass Spectrometry,* 42 CLIN. CHEM. 349 (1996); D. H. Chace et al., *Rapid Diagnosis of MCAD Deficiency: Quantitatively Analysis of Octanoylcarnitine and Other Acylcarnitines in Newborn Blood Spots by Tandem Mass Spectrometry,* 43 CLIN. CHEM. 2106 (1997); B. Wilcken et al., *Carnitine Transporter Defect Diagnosed by Newborn Screening with Electrospray Tandem Mass Spectrometry,* 138 J. PEDIATR. 581 (2001).

14. Frazier, *supra* note 11; Wilcken, *supra* note 12; A. Schulze et al., *Expanded Newborn Screening for Inborn Errors of Metabolism by Electrospray Ionization-Tandem Mass Spectrometry: Results, Outcome, and Implications,* 111 PEDIATRICS 1399 (2003); T. H. Zytkovicz et al., *Tandem Mass Spectrometric Analysis for Amino, Organic, and Fatty Acid Disorders in Newborn Dried Blood Spots: A Two-Year Summary from the New England Newborn Screening Program,* 47 CLIN. CHEM. 1945 (2001).

15. Frazier, *supra* note 11; Wilcken, *supra* note 12; Schulze, *supra* note 14; Zytkovicz, *supra* note 14.

16. K. Bhattacharya et al., *Newborn Screening May Fail to Identify Intermediate Forms of Maple Syrup Urine Disease,* 29 J. INHERIT METAB. DIS. 586 (2006).

17. S. Snyderman & C. Sansaricq, *Newborn Screening for Homocystinuria,* 48 EARLY HUM. DEV. 203 (1997).

18. J. M. G. Wilson & G. Jungner, Principles and Practice of Screening for Disease (1968).

19. National Screening Committee, Second Report of the National Screening Committee (2000).

20. Woods, *supra* note 6.

21. B. Wilcken et al., *Urine Screening for Aminoacidopathies: Is it Beneficial? Results of a Long-Term Follow-Up of Cases Detected by Screening One Millon Babies,* 97 J. PEDIATR. 492 (1980).

22. R. Pollitt, *International Perspectives on Newborn Screening,* 29 J. INHERIT METAB. DIS. 390 (2006).

23. M. S. Watson et al., *Main Report,* 8 GENET. MED. 12S (2006).

24. Guidelines of the German Ministry of Health and Social Affairs, http://www.aerzteblatt.de/v4/archiv/artikel.asp?id.

25. R. J. Pollitt & J. V. Leonard, *Prospective Surveillance Study of Medium Chain Acyl-CoA Dehydrogenase Deficiency in the UK,* 79 ARCH. DIS. CHILD. 116 (1998); C. A. Seymour et al., *Newborn Screening for Inborn Errors of Metabolism: A Systematic Review,* 1 HEALTH TECHNOL. ASSESS. (2001).

26. Wilcken, *supra* note 12; German Ministry of Health and Social Affairs, *supra* note 24; M. Pourfarzam et al., *Neonatal Screening for Medium-Chain Acyl-CoA Dehydrogenase Deficiency,* 358 LANCET 1063 (2001).

27. B. S. Andersen et al., *Medium-Chain Acyl-CoA Dehydrogenase (MCAD) Mutations Identified by MS/MS-Based Prospective Screening of Newborns Differ from those Observed in Patients with Clinical Symptoms: Identification and Characterization of a New, Prevalent Mutation that Results in Mild MCAD Deficiency,* 68 AM. J. HUM. GENET. 1408 (2001); L. Waddell et al., *Medium-Chain Acyl-CoA Dehydrogenase Deficiency: Genotype-Biochemical Phenotype Correlations,* 87 MOL. GENET. METAB. 32 (2006).

28. C. R. Roe & J. H. Ding, *Mitochondrial Fatty Acid Oxidation Disorders, in* The Metabolic and Molecular Bases of Inherited Disease 2297 (C. S. Scriver & W. S. Sly eds., 2005).

29. D. D. Koeberl et al., *Evaluation of 3-Methylcrotonyl-CoA Carboxylase Deficiency Detected by Tandem Mass Spectrometry Newborn Screening,* 26 J. INHERIT METAB. DIS. 25 (2003); K. M. Gibson et al., *3-Methylcrotonyl-Coenzyme A Carboxylase Deficiency in Amish/Mennonite Adults Identified by Detection of Increased Acylcarnitines in Blood Spots of their Children,* 132 J. PEDIATR. 519 (1998).

30. Frazier, *supra* note 11.

31. Screening, Technology and Research in Genetics, Genetic Fact Sheets for Parents: Organic Acids, http://www.newbornscreening.info/.

32. B. Wilcken, *Newborn Screening for Inborn Errors of Metabolism: Clinical Effectiveness,* 29 J. INHERIT METAB. DIS. 366 (2006).

33. S. E. Waisbren et al., *Newborn Screening Compared to Clinical Identification of Biochemical Genetic Disorders,* 25 J. INHERITED METAB. DIS. 599 (2002).

34. Frazier, *supra* note 11.

35. Schulze, *supra* note 14; E. M. Maier et al., *Population Spectrum of ACADM Genotypes Correlated to Biochemical Phenotypes in Newborn Screening for Medium-Chain Acyl-CoA Dehydrogenase Deficiency,* 25 HUM. MUTAT. 443 (2005).

36. E. Beutler, *Lysosomal Storage Diseases: Natural History and Ethical and Economic Aspect,* 87 MOLECULAR GENETETICS & METABOLISM (forthcoming 2006).

37. M. H. Gelb et al., *Direct Multiplex Assay of Enzymes in Dried Blood Spots by Tandem Mass Spectrometry for Newborn Screening of Lysosomal Disorders,* 29 J. INHERIT METAB. DIS. 397 (2006); P. J. Meikle et al., *Newborn Screening for Lysosomal Storage Disorders,* 88 MOLECULAR GENETETICS AND METABOLISM (forthcoming 2006).

38. M. L. Escolar et al., *Transplantation of Umbilical Cord Blood in Babies with Infantile Krabbe's Disease,* 352 N. ENGL. J. MED. 2069 (2005).

39. H. J. Lai et al., *The Survival Advantage of Patients with Cystic Fibrosis Diagnosed through Neonatal Screening: Evidence from the United States Cystic Fibrosis Foundation Registry Data*, 147 J. PEDIATR. (supplement 3) S57 (2005).
40. H. G. Welch & W. Burke, *Uncertainties in Genetic Testing for Chronic Disease*, 280 J. A. M. A. 1525 (1998).

Screening Newborns for Genetic Susceptibility: What's the Harm?

Nikki KERRUISH

Department of Pediatrics and Child Health, Otago University, Dunedin, New Zealand

1. INTRODUCTION

The achievements of the Human Genome Project, and subsequent developments, have been remarkable. They have led to considerable advances in our understanding of the influence of genomic variation on human diversity and susceptibility to disease, and promise great opportunities for the prediction, treatment, and prevention of common, complex diseases.[1] Many challenges remain, with one of the most significant being how to develop strategies for incorporating genetic susceptibility testing (or genomic profiling) into clinical practice.[2]

There is a lack of consensus regarding how likely this is to occur and about the expected time course for developments if it does. Some claim that the rewards of the human genome project will include "a new understanding of the genetic contribution to human disease and the development of rational strategies for minimizing or preventing disease phenotypes altogether."[3] Others are more cautious, suggesting that decades of epidemiological study and clinical evaluation of interventions will be required.[4] Some are skeptical that genomics will ever revolutionize the way in which common diseases are identified or prevented, with their doubts stemming from issues such as incomplete penetrance of genotypes for common diseases, the limited ability to tailor treatment to genotypes, and the low magnitude of risks conferred by various genotypes for the population at large.[5]

We cannot be certain about how useful genomics will prove to be in the clinical setting, but we can say that there are obvious advantages to using a pre-existing screening infrastructure if susceptibility testing is to be implemented at a population level. This point has not gone unrecognized, with one suggestion that has attracted attention being to utilize samples taken for existing newborn screening programmes for genomic profiling.[6] It therefore seems prudent to consider the ethical consequences of integrating genetic susceptibility testing into clinical practices such as newborn screening.

This chapter aims first to describe what is meant by susceptibility testing and then to briefly discuss some of the practical or scientific issues that need to be addressed. It will then highlight some of the ethical and social issues that may arise when incorporating susceptibility testing into newborn screening, and in particular will discuss the potential for harmful effects upon children and families that may arise from a combination of factors related to the tests and features of the newborn period.

2.   GENETIC SUSCEPTIBILITY TESTING

Genetic susceptibility testing: "Testing for DNA sequence variation (SNPs) associated with increased or decreased risk of disease."

Genomic profiling: "Concurrent detection of multiple gene variants (SNPs) that have been associated with greater risk or predisposition to a particular disease or condition."[7]

About 99.8% of human DNA sequences are identical across the population. Genomic profiling focuses on the 0.2% that is variable. These variations are termed single nucleotide polymorphisms (SNPs) and refer to single nucleotide or single letter spelling differences between the DNA of different individuals. SNPs make up about 90% of all human genetic variation and occur every 100 to 300 bases along the 3-billion-base human genome. This means that the human genome has about 10 million polymorphisms, defined as genetic variants in which the minor gene forms occur at least once out of every 100 forms, and any two unrelated humans will have millions of genetic differences that contribute to making them look and behave differently.[8] These variations may predispose or protect individuals from developing various disorders and affect how people respond to disease; environmental insults such as bacteria, viruses, toxins, and chemicals; and drugs and other therapies. They also contribute to biological variation such as height and metabolism, and some are thought to have no effect (2006).

Many of the conditions in which SNPs play important roles are common, multifactorial disorders, such as diabetes, cardiovascular disease and obesity. The difficulty is that although gene variants *predispose* individuals to common diseases, they do not *cause* disease in isolation, rather operating in a highly complex manner in combination with other gene variants and with the environment.[9] So, whereas a positive newborn screening test for PKU means that the biochemical disorder is already present and the disease will develop rapidly without treatment, a positive susceptibility test gives an individual information about their personal risk of developing a disease sometime in the future. This type of information is derived from population genetic studies and is usually presented in terms of a probability estimate or odds ratio, so for individuals some uncertainty remains as to whether they will develop the condition, and if so, when.[10]

3.   TYPE 1 DIABETES: A DISEASE MODEL

Type 1 diabetes (T1D) represents a useful model on which to base discussions concerning the potential use of genetic susceptibility testing in clinical practice. It is one of the most common chronic childhood diseases, with a rising incidence (3-4% per year in most developed countries), particularly in the 0-4 yr age group.[11] At present, development of the disease necessitates life-long adherence to a difficult therapeutic regime that is only partially effective in preventing acute and chronic complications.[12] These facts, as well as the existence of a long period of latent autoimmunity preceding the onset of clinical diabetes, make the possibility of disease prevention an attractive and potentially achievable goal.[13]

T1D is representative of the type of conditions that may be screened for using genetic susceptibility tests in that it is a multifactorial condition with disease development dependant upon an undetermined number of genetic and environmental factors.[14] The chief genetic determinant of susceptibility to diabetes lies within the class II region of the major histocompatibility complex on chromosome 6.[15] More than 90% of patients who develop T1D have either the DR3 and/or the DR4 allele of the HLA-DRB1 gene, whereas fewer than 40% of healthy controls have these alleles. Depending on the population, people homozygous for the high-risk DR4 allele have a 10-15 fold increased risk of T1D and people heterozygous for the DR3/DR4 alleles a 20-30 fold increased risk.[16] Of the HLA-DRB1 *04 subtypes, *0401, *0402, *0404, *0405 confer the highest relative risk to T1D, whereas *0403, *0406 and *0408 confer protection to T1D.[17]

This concentration of the genetic risk makes general population screening for higher risk individuals economically feasible using current technology, and analytical costs are likely to decline as new techniques emerge.[18] Population screening for genetic susceptibility to T1D does not form part of current clinical practice but is the object of several longitudinal prospective studies,[19] and more recently, a larger prospective study involving a consortium of six centers in the US, Scandinavia and Europe.[20] There are currently no preventative treatments, but these studies follow children in order to determine whether and when the child develops auto antibodies (preclinical disease) or overt diabetes. The explicit aim of the studies is to further elucidate the natural history of T1D, to identify environmental exposures that may trigger autoimmunity and ultimately to test interventions that may prevent the disease. T1D is one of the first of many diseases with complex genetic and environmental determinants that is being studied in this manner, but the characterization of at risk groups by genotype with the view to instituting preventative measures could equally apply to other common multifactorial conditions, such as asthma and obesity.

4.   CHALLENGES INVOLVED IN INTEGRATING SUSCEPTIBILITY TESTING INTO CLINICAL PRACTICE

Large prospective studies such as those investigating the pathogenesis of T1D provide us with much information and many opportunities concerning the use of susceptibility testing in the newborn period. The remainder of this chapter will discuss some of the challenges that need to be addressed before it will be possible to translate advances in the field of genomic medicine into health benefits. It will refer specifically to information that can be derived from prospective T1D studies to illustrate various points and will highlight opportunities to gather more empirical evidence.

The challenges are numerous, and they are also varied, spanning scientific and practical issues as well as ethical, legal and social issues. In practice, these areas are closely intertwined and difficult to discuss in isolation, but for the sake of clarity they will be considered separately here. Scientific and practical issues are not the main focus of this discussion and will be covered only briefly, with reference to work by those who have discussed these issues in considerable depth. This will leave greater scope to discuss ethical and social issues in more detail, and in particular to more fully articulate those that are specific to testing in the newborn period.

5.   SCIENTIFIC AND PRACTICAL ISSUES

All medical tests vary in how well they are able to predict a particular outcome, and genetic susceptibility tests are no different in this respect. The need to evaluate genetic tests in population-based settings before their use in clinical practice has been recognized for some time[21] and an approach developed that can help to determine the potential value of the test in patient care. This approach is called ACCE (Analytic validity; Clinical validity; Clinical utility; and Ethical, legal, and social implications).[22] The system is well developed and consists of 44 targeted questions intended to determine what is currently known about the test and disorder in question, as well as to identify gaps in current knowledge.[23]

Analytic validity refers to the accuracy of the test in identifying the genotype of interest. This encompasses analytic sensitivity and specificity as well as issues related to laboratory quality control.[24] *Clinical validity* refers to the accuracy and reliability with which a test detects or predicts a particular clinical outcome and includes the detection rate and false positive rate as well as penetrance.[25] All of these factors can be affected by the setting in which the test is performed (e.g., population screening versus clinical diagnosis) and the frequency of the mutation in the population.[26] *Clinical utility* refers to the usefulness of the test and the value of the information to the person being tested.[27] Measuring clinical utility can be difficult as it requires evaluation of the risks and benefits of testing. This may include knowledge of the natural history of the disorder; availability, uptake and effectiveness of interventions, impact of interventions on health outcomes; cost-effectiveness; and social acceptability.[28]

For most genetic susceptibility tests, there are currently significant gaps in our knowledge concerning analytic validity, clinical validity and utility, or the test does not rate well enough in these areas to be considered for clinical use. For example, in the case of T1D, a HuGENet review concerning the HLA-DQ locus noted the relatively low sensitivity and specificity estimates of high-risk alleles in the general population and the current lack of an effective preventive intervention. It also articulated the need for more information concerning population-based risk factor specific incidence rates in all ethnic groups as well as the need to consider genetic counselling services and genetic education for T1D families and health professionals.[29]

In order to obtain good information about analytic and clinical validity and utility, high quality (and expensive) biomedical research will be required, and a clear vision for how this research should progress has been developed.[30] This seminal paper provides "an overview of the broad landscape of scientific opportunity" and suggests a varied range of scientific endeavours. These include collection of large-scale genomic data sets such as the HapMap project[31] and longitudinal population cohort studies designed to identify genetic and environmental contributors to health and to assess the effect of individual gene variants on disease risk.[32]

6.   THE SPECIAL STATUS OF THE NEWBORN PERIOD

There may be ethical and social concerns related to testing people of any age for genetic susceptibility to disease. These include the potential for stigmatization, discrimination and concerns related to how to achieve a balance between appropriate access to clinically important results and privacy.[33] However, there are particular features of the newborn period that necessitate consideration of some additional ethical and social issues. These features of the newborn period include:

- the newborn baby's lack of personal capacity to consent to or decline screening,
- the newborn period represents a critical phase of infant-parent bonding
- external influences, including the psychological or emotional state of the parents, can have a profound and permanent effect on child development. This issue continues into early childhood, with the first two years of life being considered the most important from a developmental perspective.[34]

With respect to these features of the newborn period, this chapter will now discuss two significant ethical issues, namely consent for testing and potential harmful effects of testing.

7.   CONSENT FOR NEWBORN SCREENING

For many multifactorial conditions, the newborn period would be the ideal time to test for genetic susceptibility as elements of the early childhood environment are probably important factors in the later development of disease. For instance, environmental factors that have been postulated as having a role in the pathogenesis of T1D include early exposure to cow's milk and early childhood, or even in utero, enteroviral infection.[35] If we are to affect the incidence of these diseases, preventative measures will most likely be required in the early years. However, testing soon after birth necessitates consideration of issues related to proxy consent and the evolving autonomy of the child.

Informed consent is one of the basic elements of medical ethics and the professional-patient relationship. The doctrine of informed consent reminds us to respect persons by fully and accurately providing information relevant to them exercising their decision-making rights.[36] Belief in this doctrine has led to attempts to adapt the concept to newborn testing, with many believing that a baby's parents should play the key role in such a process and give consent by proxy. Proxy consent seeks to protect the best interests of the child, and as this is also the aim of most parents, the concept generally appears to work well in pediatric practice.

7.1   *Proxy Consent: Practicalities and "Standard Newborn Screening"*

However, the issue of informed consent for even standard newborn screening (e.g. the Guthrie test for PKU) has provoked debate, as is reflected in the considerable variation at both policy level and in the practical delivery of programmes. For instance, the WHO Guidelines[37] consider newborn screening to be sufficiently important to override parental refusal, stating that "newborn screening should be mandatory and free of charge if early diagnosis and treatment will benefit the newborn." Despite this, there appears to have been a shift towards informed parental choice in newborn screening although it is still mandated in some US states.[38] The practice of mandated newborn screening has been both praised and criticized.[39] Those in favour generally cite the risk of harm that could be avoided through screening and early diagnosis[40] whereas those against suggest consent helps foster the idea of "partnership" between parents and professionals in providing care for a child, is a symbol of respect for the family and educates parents about the value and purpose of screening.[41]

Overall, consent practices for newborn screening are poorly described and probably vary markedly within, and between, different jurisdictions. It is likely that many consent processes operate on an opt-out basis, whereby parental consent is assumed if no objections are voiced.[42] However, most current newborn screening programmes, even if testing is mandatory, articulate a commitment to informing parents and provide websites and information sheets concerning the testing process and conditions tested for.[43] In practice, fully informed choice may be difficult to achieve due to the volume of information presented during pregnancy and in the immediate postnatal period. A recent survey of women in Australia showed that although they were aware of newborn screening, they did not consider they had comprehensive knowledge of the tests.[44]

These difficulties may be exaggerated if genetic susceptibility tests are included in newborn screening protocols. For example, in the case of T1D, an assessment of maternal understanding of infant T1D risk at 4 months post-notification found that only 62% correctly estimated their child's genetic risk, with 24% underestimating.[45] These difficulties would be further accentuated if newborn genomic profiling were undertaken. For instance, it would seem impossible to obtain specific consent for the hundreds or even thousands of conditions that profiling may reveal, but it is not clear whether a more generic consent is appropriate, or if so, what form it should take. Comprehending the information is one important element of informed consent, but how one actually weighs up the potential harms and benefits of testing in order to decide whether or not to consent is another. The complexity of the information generated may require an appreciation not only of particular diseases and their treatments, but also some understanding of epidemiology, population and individual risk, and statistics.[46] At current levels of scientific knowledge, there is disagreement amongst both professionals and lay people regarding the acceptability and utility of testing,[47] and it may be onerous for parents to decide upon an appropriate course for their newborn.

7.2   *Proxy Consent: Ethical Theory and Evolving Autonomy*

Despite its widespread use, there are some fairly obvious inherent ambiguities in the concept of proxy consent, in that consent generally expresses something of oneself. In other words, a person who consents does so on the basis of their own unique personal beliefs and values,[48] but this cannot be so in newborn screening as it is not possible to accurately gauge what a child's future beliefs will be. While it is clear that a practical solution to the newborn's incapacity to consent is required, and it also seems reasonable to suggest that the most appropriate approach is parental proxy consent, we need to be mindful of this significant theoretical difference between proxy consent and individual informed consent.

Some authors,[49] recognizing the unique moral status of children, have described children as possessing "rights in trust," which have properties in common with autonomy-based rights of adults. As young children are unable to exercise these rights until they are deemed competent, Feinberg suggests that parents have an obligation not to let their values interfere with respecting what their child may want for his or her future self. He argues that children's future autonomy should be maximized until they are able to make decisions for themselves, thus protecting their right to an "open future."[50]

Feinberg would likely suggest that parents should not refuse standard newborn screening for reasons including that failure to detect PKU, for example, would seriously impact on their child's later ability to behave autonomously. However, it is less clear how this standard applies to genetic susceptibility testing. Results from genetic susceptibility tests are not only probabilistic but they are also predictive. They do not demonstrate that a disorder is present but that it may occur some time later in childhood. Most official policies concerning predictive genetic testing strongly advise against testing children for a disease in which surveillance, pre-emptive or definitive medical treatment is not available in childhood.[51] This approach protects the child's future autonomy to self-determine whether or not to be tested and does not violate the future adult's right not to know.

However, alternative arguments that children's best interests should not be considered in narrow, medical terms but according to a broader definition including biological, social and psychosocial elements, have challenged the prohibition on predictive genetic testing in childhood. Self-knowledge (including genetic test results), it is argued, can promote more autonomous decision-making and allow better psychosocial adjustment.[52] In other words, growing up knowing that one is at risk of developing a disorder could be viewed as enhancing a child's developing autonomy, rather than constraining it. While this argument has much merit when considering fully penetrant, monogenic disorders, the same does not necessarily follow for genomic profiling as it may be difficult, and potentially harmful, for parents or children to try to plan their lives on the basis of tests with variable predictive value. Clearly, if accuracy of testing improves, as some predict, this statement will become less relevant.

There is little empirical evidence concerning the attitudes of parents and children toward these issues, although a qualitative research report from the US suggests that parents believe that they, and not professionals, should be the final arbiters of what their child is tested for.[53] Children's voices are generally absent from the debate, and as some children who have undergone newborn testing for specific genetic susceptibilities such as T1D reach an age where they can discuss their views, it will be important for them to be included.

8. POTENTIAL HARMFUL EFFECTS OF GENETIC SUSCEPTIBILITY TESTING

As noted previously, most published guidelines concerning predictive genetic testing of children[54] oppose testing in the absence of known medical benefits. The reasoning behind this prohibitive stance relates not only to an attempt to protect the evolving autonomy of the child, but also because of concerns about psychosocial sequelae that genetic testing in childhood may generate.

What harm could come to children who undergo genomic profiling? In order to answer this question, it is necessary to consider what is meant by harm in this context, and which children it may apply to. Many of the adverse effects of this type of testing are likely to be indirect, psychosocial and perhaps even difficult to define. If we are interested in knowing what may happen as a result of genomic profiling, then we must consider a wide range of potential harms or indeed anything that impacts on the "welfare of the child." It is also important to consider the fact that the majority of gene-positive children will not actually develop the disease to which they are genetically predisposed, but may still be subject to some of the harmful effects.

8.1   Potential Physical Harms

The physical harms associated with the heel prick blood test for newborn screening are minimal. However, there are other potential physical discomforts associated with this type of testing. Firstly, if a child tests positive for a particular susceptibility gene, they will presumably require some sort of surveillance throughout childhood and a preventative measure may also be suggested. For instance, children who test positive for susceptibility alleles for T1D require 3-6 monthly blood tests for autoantibody screening, and some may be enrolled in prevention trials.

The potential harms associated with surveillance or preventative measures are largely predictable and can be carefully assessed in trials prior to any proposed clinical introduction of newborn genomic profiling. However, other physical harms may be less predictable, and more dependant on the reaction of individual parents to their child's test result. For example, a recent study of mothers of children at genetic risk for T1D has shown that many of them alter the way they treat their child despite there being no current medical recommendation to do this. This included monitoring behaviours (e.g. checking blood glucose), and some mothers modified the child's diet and limited physical activity.[55] In the absence of definitive preventative measures, the potential for harm may be increased if the natural parental urge to protect one's child drives a search for preventative or therapeutic strategies. For example, some parents in T1D studies are known to have altered their child's milk consumption from standard cow's milk to milk containing a different protein, because of largely unsubstantiated claims that the former triggers diabetes whereas the latter does not.[56] While this strategy is unlikely to be harmful, other similar interventions may not be so innocuous, particularly when one considers the permanent neurodevelopmental changes that may occur in relation to environmental influences in early childhood. Studies on screening for hypercholesterolemia have reported that some parents restrict their child's diet to the extent that they become malnourished.[57] Similarly, identification of a genetic predisposition to haemachromatosis in a child could lead to unnecessary restriction of iron intake with adverse neurodevelopmental effects. In the case of susceptibility testing, these negative effects may accrue in large numbers of children who were never destined to develop the condition to which they are "susceptible." One might argue that these reactions could be remediated by educative measures but this would require considerable resource allocation and may still be only partially effective. Even with optimal counselling services, concepts of risk are difficult to convey, and reactions depend upon a complex interplay of individual characteristics.[58] This is an important area for future research, and strategies to address these issues should be incorporated into prospective cohort studies wherever possible.

8.2   *Potential Psychological Harms*

Two important aspects of the newborn period that may be relevant to discussions of potential harmful effects of newborn genetic susceptibility testing have been highlighted. These are that the first few months of life represent a critical phase of infant-parent bonding and that external influences, including the psychological or emotional state of the parents, can have a profound and permanent effect on child development.[59]

In relation to the first of these issues, it is worth reviewing the literature concerning newborn screening for other disorders. Reports from Wales, where newborn screening for Duchenne Muscular Dystrophy (DMD) occurs in some regions, have found no negative effect of newborn screening on the early mother-baby relationship. In particular, there was no evidence, of either rejection or overprotection of infants diagnosed presymptomatically through screening during the first year of life.[60] Parents may prefer early diagnosis through screening, even for untreatable diseases such as DMD, with the declared advantage being that they can prepare themselves emotionally and practically and consider their reproductive options.[61]

Evidence from the literature concerning newborn screening for cystic fibrosis (CF) is less clear cut. A study comparing the strength of overprotective child rearing attitudes of 29 mothers whose children were screened (13 had symptomatic children and 16 asymptomatic children) with the attitudes of 29 mothers whose children were diagnosed after the onset of symptoms indicated that newborn screening had not increased a mother's tendency to overprotect her child with CF, and in some cases the tendency had decreased. The authors also noted that delays in diagnosis when screening was not conducted usually caused mothers considerable personal distress.[62] The reduction of diagnostic delay is generally considered to be one of the major psychosocial advantages of newborn screening for CF.[63]

However, a study describing parents' attitudes toward newborn screening for CF revealed that a minority of mothers reported experiencing difficulties with infant bonding in relation to the screening process. Some of these mothers acknowledged temporary rejection of their babies during the period of uncertainty between initial positive IRT screen and substantive diagnosis.[64] A more recent study has also highlighted the emotional distress parents experience during this period of diagnostic uncertainty.[65]

There are still some gaps in our knowledge concerning whether presymptomatic diagnosis of monogenic disorders through newborn screening affects parent-child bonding, and if so, what implications this has for the child's future. An area that is particularly under-researched is the impact of presymptomatic diagnosis or genetic risk information upon fathers and the bonding process between fathers and their children. Newborn screening for genetic susceptibility to multifactorial disorders differs from screening for DMD or CF, as it cannot provide the "benefit of certainty" that is associated with tests for monogenic disease.[66] Rather, it highlights a specific level of *uncertainty*: parents must learn to live with the knowledge that their child might develop a condition, of uncertain clinical severity, at some point during childhood. The reported difficulties of parents dealing with much briefer periods of uncertainty in relation to cystic fibrosis screening[67] may be relevant in this situation and should be investigated further. In addition, any physical, emotional or practical preparation that parents make on the basis of genetic risk information from susceptibility tests may be pointless if the child never develops the condition (T1D, for example) as the majority of children with increased risk genotypes will not.. Further investigation of whether or not newborn genetic susceptibility screening causes alteration in the early stage of bonding, and whether or not this can have lasting effects on the emerging relationship between mother and child, are required.

The second relevant feature of the newborn period and early childhood is the capacity for environmental influences, including the psychological or emotional state of the parents, to have a profound and permanent effect on child development. Again, the *uncertainty* associated with genetic susceptibility testing may be of relevance. Possessing knowledge that a child might develop a disease such as T1D at some stage in the future can be viewed in competing ways. Some parents may become distressed, particularly if no preventative measure is available; others may feel empowered by the knowledge and their opportunity to detect potential problems early and potentially minimize morbidity. Some parents may view their "at-risk" child as actually being ill or "uniquely vulnerable" and may over-attribute symptoms to the perceived risk status.[68] A parent's belief that their child is in some way vulnerable, or particularly susceptible to illness, can potentially have adverse effects upon the child's development.[69] This was first termed the "vulnerable child syndrome" by Green and Solnit in 1964, reporting on a cohort of families whose children had suffered life-threatening illnesses in infancy and then completely recovered.[70] The authors observed that the parents, particularly the mothers, continued to be anxious about their child's health, and feared the child may die. It appeared that the parents' perception of their child as being uniquely vulnerable led to difficulties in parent–child interaction. In particular, the parents overprotected the child, were unable to set age-appropriate limits and displayed excessive concerns about their child's health in medical settings. The children, apparently responding to their parents' expectations of vulnerability, showed exaggerated separation anxiety, sleep disorders, discipline problems, school underachievement and distorted perceptions of their own health.[71]

Since the original description, aspects of the vulnerable child syndrome have been described in relation to many other conditions and disorders. For instance, parents whose baby had a false positive newborn screening test for phenylketonuria continued at times to fear that their child would be developmentally delayed.[72] Forty percent of parents of children with innocent heart murmurs imposed physical and psychological restrictions on their children despite there being no evidence of organic cardiac disease. The authors of this report concluded that disability from "cardiac non-disease" in childhood was greater than that due to actual heart disease.[73] More recently, a study has suggested that higher parental perception of child vulnerability is correlated with a worse developmental outcome in premature infants at 1-year adjusted age.[74] Although not all these families displayed the florid behavioral problems described in Green and Solnit's original report, these studies provide evidence that heightened parental perceptions of a child's vulnerability may contribute to long-term developmental problems.

Empirical evidence concerning parental reaction to newborn susceptibility testing is gradually accruing, particularly in relation to newborn T1D screening. It appears that in fact very few parents are significantly distressed by this type of genetic knowledge. For example, Bennett Johnson et al in Florida studied 435 mothers of infants with increased genetic risk of T1D at 4 months and 1 year post-risk notification and found that for most mothers, this type of newborn genetic screening was not associated with significantly elevated maternal anxiety, and that anxiety further dissipated over time. Some mothers (for example, Hispanic mothers and those with infants sub-classified as extremely high-risk) did experience more anxiety than pregnant or working women comparison groups, and this merits further investigation.[75] Yu et al. in Colorado studied 23 mothers of infants at high genetic risk of T1D, and 65 mothers of low-risk infants, using the Parenting Stress Index (PSI) at baseline (5-7 weeks postpartum) and 4-5 months after risk notification. They did not find a statistically significant association between infant genetic risk status and change in maternal scores on the PSI.[76] Research from the Department of Pediatrics and Child Health at the University of Otago involving a cohort of mothers and their babies with increased genetic risk of T1D concurs with other investigators' conclusions that there is no clinically significant psychosocial disturbance as measured on standard rating scales. An important additional finding was that there was no evidence that mothers perceived their babies with increased genetic risk of diabetes to be any more vulnerable or fragile than did mothers of low-risk babies. This was assessed using a questionnaire specifically designed to measure maternal perceptions of vulnerability in very young infants[77] and suggests that the "vulnerable child syndrome" is unlikely to occur in relation to genetic risk status for T1D.[78]

Despite this reassuring data, when asked to rate their own degree of concern about their baby's genetic risk, mothers of babies at increased genetic risk reported significantly higher levels than mothers of babies with low or unknown genetic risk. Of course, this is not a surprising result: it seems unrealistic to think there will be no difference in psychosocial reaction between mothers of genetically susceptible or low-risk babies. In fact, one of the aims of any newborn screening programme involving genetic susceptibility testing must be to create some degree of heightened awareness of their child's health risks among parents so that they participate in surveillance and/or preventative measures. It would clearly be wrong to overburden parents and create problems akin to the vulnerable child syndrome, but just how much parental concern should we aim to generate?

Determining precisely what an appropriate or acceptable parental response constitutes, and how best to achieve this in practice, are challenges for the future.

Finally, it is not only the children who test positive for an increased risk genotype who are at risk of potential harm through susceptibility testing. Having a low-risk genotype for a multifactorial disease does not eliminate the possibility of the condition developing. For the susceptibility tests currently used to screen for T1D, it simply means that the risk is low (less than 1 in 1500). It is imperative that parents of these children are not falsely reassured and still recognize the symptoms of developing illness were they to develop.

9.   CONCLUSIONS AND IMPLICATIONS FOR FUTURE RESEARCH

Many challenges have been recognized since the completion of the Human Genome Project, with perhaps one of the most significant being how genetic susceptibility testing (or genomic profiling) might be integrated into medical practices, such as newborn screening.

This chapter has briefly discussed scientific or practical matters such as analytic validity and clinical validity and utility. These important issues can only be adequately addressed with high-quality biomedical research including large well designed epidemiological studies that aim to elucidate the complex pathogenesis of multifactorial disorders. In the case of T1D, these are underway, and a clear vision for how this research in general should progress has been developed,[79] again including the use of longitudinal population cohort studies designed to identify genetic and environmental contributors to health and to assess the effect of individual gene variants on disease risk.[80] As these studies proceed, interventions designed to prevent disease onset may become available and will also require careful testing in longitudinal intervention studies.

It is imperative that empirical research into ethical and psychosocial issues is incorporated into the studies detailed above. For some of the issues raised in this chapter, it is unlikely that further theoretical advances in the debate can be made without empirical data. In order to develop a more sophisticated appreciation of all the harms and benefits associated with genetic susceptibility testing, particularly in the newborn period, research into ethical and psychosocial issues needs to be a fundamental part of the design of epidemiological studies. This would serve to encourage formulation of the robust and detailed proposals that will be required to address the complex issues that are likely to arise when conveying genetic risk information to parents of newborn babies. Particular issues that merit attention include the mechanisms involved in obtaining consent for testing and how this impacts on the evolving autonomy of the child. In addition, the potential for harmful physical and psychosocial effects requires further evaluation, and it will be important to try to determine precisely what we consider to be an acceptable level of parental concern in response to genetic susceptibility information, and how best to achieve this in practice.

The ethical and psychosocial issues highlighted cannot be ignored if genetic susceptibility testing is to be utilized in the newborn period. Addressing the issues may well be challenging, but as scientific research continues there is time to conduct high quality investigation in these areas. The research strategies required will vary but should include non-empirical philosophical research as well as empirical evaluation by multidisciplinary teams using a combination of quantitative and qualitative methodologies. In the meantime, ethical and psychosocial issues that are currently under-researched, and can potentially be addressed satisfactorily, should not be exaggerated, as this may deter people from using future genetic services and deprive them of significant clinical benefits.

REFERENCES

1. F. S. Collins & V. A. Mckusick, *Implications of the Human Genome Project for Medical Science*, 285 J. A. M. A. 540 (2001).
2. C. A. Moore et al., *From Genetics to Genomics: Using Gene-Based Medicine to Prevent Disease and Promote Health in Children*, 29 SEMIN. PERINATOL. 135 (2005).
3. F. S. Collins, *Shattuck Lecture – Medical and Societal Consequences of the Human Genome Project*, 341 N. ENGL. J. MED. 28 (1999).
4. S. B. Haga et al., *Genomic Profiling to Promote a Healthy Lifestyle: Not Ready for Prime Time*, 34 NAT. GENET. 347 (2003).
5. N. A. Holtzman & T. M. Marteau, *Will Genetics Revolutionize Medicine?* 343 N. ENGL. J. MED. 141 (2000); A. O. M. Wilkie, *Genetic Prediction: What Are the Limits?* 32 STUD. HIST. PHIL. BIOL. AND BIOMED. SCI. 619 (2001).
6. B. Almond, *Genetic Profiling of Newborns: Ethical and Social Issues*, 7 NAT. REV. GENET. 67 (2006); Human Genetics Commission, Profiling the Newborn: A prospective Gene Technology? (2005), http://www.hgc.gov.uk/Client/news_ item.asp?NewsId=38.
7. Haga, *supra* note 4.
8. D. B. Goldstein & G. L. Cavalleri, *Genomics: Understanding Human Diversity*, 437 NATURE 1241 (2005).
9. *Id.*
10. N. J. Kerruish & S. P. Robertson, *Newborn Screening: New Developments, New Dilemmas*, 31 J. MED. ETHICS 393 (2005).
11. EURODIAB ACE Study Group, *Variation and Trends in Incidence of Childhood Diabetes in Europe*, 355 LANCET 873 (2000); P. L. Campbell-Stokes & B. J. Taylor, *Prospective Incidence Study of Diabetes Mellitus in New Zealand Children Aged 0 to 14 Years*, 48 DIABETOLOGIA 643 (2005).
12. D. Devendra et al., *Type 1 Diabetes: Recent Developments*, 328 B. M. J. 750 (2004).
13. R. F. Vogt et al., *Newborn Screening and Type 1 Diabetes: Historical Perspective and Current Activities at the CDC Division of Laboratory Sciences*, 5 DIABETES TECHNOL. THER. 1017 (2003).
14. R. Buzzetti et al., *Dissecting the Genetics of Type 1 Diabetes: Relevance for Familial Clustering and Differences in Incidence*, 14 DIABETES METAB. REV. 111 (1998).
15. J. A. Todd, *From Genome to Aetiology in a Multifactorial Disease, Type 1 Diabetes*, 21 BIOESSAYS 164 (1999).
16. J. X. She, *Susceptibility to Type 1 Diabetes: HLA-DQ and DR Revisited*, 17 IMMUNOL. TODAY 323 (1996).
17. *Id.*
18. Vogt, *supra* note 13.
19. M. Greener, *PANDA Identifies Babies at Risk of Developing Type 1 Diabetes*, 6 MOL. MED. TODAY 3 (2000); M. Rewers et al., *Newborn Screening for HLA Markers Associated with IDDM: Diabetes Autoimmunity Study in the Young (DAISY)*, 39 DIABETOLOGIA 807 (1996); K. Sadeharju et al., *Enterovirus Infections As a Risk Factor for Type 1 Diabetes: Virus Analyses in a Dietary*

Intervention Trial, 132 CLIN. EXP. IMMUNOL. 271 (2003); L. Berzina et al., *Newborn Screening for High-Risk Human Leukocyte Antigen Markers Associated with Insulin-Dependent Diabetes Mellitus: The ABIS Study,* 958 ANN. N. Y. ACAD. SCI. 312 (2002).

20. The Environmental Determinants of Diabetes in the Young (TEDDY) Study Website, http://teddy.epi.usf.edu/.

21. Moore, *supra* note 2; N. Holtzman & M. Watson, *Promoting Safe and Effective Genetic Testing in the United States, Final Report of the Task Force on Genetic Testing,* 2 J. CHILD FAM. NURS. 388 (1999); Secretary's Advisory Committee on Genetic Testing, Enhancing the Oversight of Genetic Tests: Recommendations of the SACGT (2001), http://www4.od.nih.gov/oba/sacgt/reports/oversight _report.pdf.

22. J. Haddow & G. Palomaki, *ACCE a Model Process for Evaluating Data on Emerging Genetic Tests, in* Human Genome Epidemiology (Y. Khoury et al. eds., 2004).

23. Moore, *supra* note 2.

24. W. Burke et al., *Genetic Test Evaluation: Information Needs of Clinicians, Policy Makers, and the Public,* 156 AM. J. EPIDEMIOL. 311 (2002).

25. *Id.*

26. Moore, *supra* note 2.

27. W. Burke et al., *Categorizing Genetic Tests to Identify their Ethical, Legal, and Social Implications,* 106 AM. J. MED. GENET. 233 (2001).

28. Moore, *supra* note 2.

29. CDC Website, Genomics and Disease Prevention, http://www.cdc.gov/genomics /hugenet/reviews/diabetes.htm.

30. F. S. Collins et al., *A Vision for the Future of Genomics Research,* 422 NATURE 835 (2003).

31. G. A. Thorisson et al., *The International HapMap Project Web Site,* 15 GENOME RES. 1592 (2005).

32. Collins, *supra* note 30.

33. Almond, *supra* note 6; P. A. Baird, *Identification of Genetic Susceptibility to Common Diseases: The Case for Regulation,* 45 PERSPECT. BIOL. MED. 516 (2002); Human Genetics Commission, *supra* note 6.

34. S. J. Webb et al., *Mechanisms of Postnatal Neurobiological Development: Implications for Human Development,* 19 DEV. NEUROPSYCHOL. 147 (2001).

35. H. K. Akerblom et al., *Environmental Factors in the Etiology of Type 1 Diabetes,* 115 AM. J. MED. GENET. 18 (2002); Sadeharju, *supra* note 19.

36. Committee on Bioethics, American Academy of Pediatrics, *Informed Consent, Parental Permission, and Assent in Pediatric Practice,* 95 PEDIATRICS 314 (1995).

37. Anonymous, Proposed International Guidelines on Ethical Issues in Medical Genetics and Genetic Services, WHO/HGN/GL/ETH/98 1, (2000).

38. Serving the family from birth to the medical home, *Newborn Screening: A Blueprint for the Future – A Call for a National Agenda on State Newborn Screening Programs,* 106 PEDIATRICS 389 (2000).

39. A. Newson, *Should Parental Refusals of Newborn Screening be Respected?,* 15 CAMB. Q. HEALTH ETHICS 135 (2006).

40. R. R. Faden et al., *Parental Rights, Child Welfare, and Public Health: The Case of PKU Screening*, 72 AM. J. PUB. HEALTH 1396 (1982).
41. A. Clarke, *Newborn Screening*, in Genetics, Society and Clinical Practice (A. Clarke & P. Harper eds., 1997); L. Friedman Ross, *Genetic Testing of Children: Who Should Consent?*, in A Companion to Genethics (J. Harris & J. Burley eds., 2002).
42. M. C. Huang et al., *Parental Consent for Newborn Screening in Southern Taiwan*, 31 J. MED. ETHICS 621 (2005).
43. NSW Screening Programme, http://www.chw.edu.au/prof/services/newborn/.
44. A. Davey et al., *New Mothers' Awareness of Newborn Screening, and Their Attitudes to the Retention and Use of Screening Samples for Research Purposes*, GENOMICS, SOCIETY AND POLICY 1 (2005).
45. S. K. Carmichael et al., *Prospective Assessment in Newborns of Diabetes Autoimmunity (PANDA): Maternal Understanding of Infant Diabetes Risk*, 5 GENET. MED. 77 (2003).
46. Almond, *supra* note 6.
47. Collins, *supra* note 3; N. A. Holtzman & T. M. Marteau, *Will Genetics Revolutionize Medicine?* 343 N. ENGL. J. MED. 141 (2000); Wilkie, *supra* note 5.
48. Committee on Bioethics, American Academy of Pediatrics, *supra* note 36.
49. J. Feinberg, *The Child's Right to an Open Future*, in Whose Child? Children's Rights, Parental Authority, and State Power (W. Aiken & H. Lafollette eds., 1980).
50. *Id.*
51. A. Clarke, *The Genetic Testing of Children. Working Party of the Clinical Genetics Society (UK)*, 31 J. MED. GENET. 785 (1994); Anonymous, *Points to Consider: Ethical, Legal, and Psychosocial Implications of Genetic Testing in Children and Adolescents. American Society of Human Genetics Board of Directors, American College of Medical Genetics Board of Director*, 57 AM. J. HUM. GENET. 1233 (1995); Human Genetics Society of Australasia, Predictive Testing in Children and Adolescents (2003), http://www.hgsa.com.au/policy/ptca.html.
52. S. Robertson & J. Savulescu, *Is There a Case in Favour of Predictive Genetic Testing in Young Children?*, 15 BIOETHICS 26 (2001).
53. E. Campbell & L. F. Ross, *Parental Attitudes Regarding Newborn Screening of PKU and DMD*, 120 AM. J. MED. GENET. 209 (2003).
54. Clarke, *supra* note 51; Anonymous, *supra* note 51; Human Genetics Society of Australasia, *supra* note 51.
55. A. E. Baughcum et al., *Maternal Efforts to Prevent Type 1 Diabetes in At-Risk Children*, 28 DIABETES CARE 916 (2005).
56. Kerruish, *supra* note 10.
57. S. B. Hulley & T. B. Newman, *Position Statement: Cholesterol Screening in Children Is not Indicated, even with Positive Family History*, 11 J. AM. COLL. NUTR. (supplement) 20S (1992).
58. S. Michie & T. Marteau, *Predictive Genetic Testing in Children: The Need for Psychological Research*, in The Genetic Testing of Children (A. Clarke ed., 1998).
59. Webb, *supra* note 34.

60. E. P. Parsons et al., *Newborn Screening for Duchenne Muscular Dystrophy: A Psychosocial Study,* 86 ARCH. DIS. CHILD FETAL. NEONAT. ED. F91 (2002).
61. Parsons, *supra* note 60; D. M. Hall & J. M. Michel, *Screening in Infancy,* 72 ARCH. DIS. CHILD 93 (1995).
62. C. Boland & N. L. Thompson, *Effects of Newborn Screening of Cystic Fibrosis on Reported Maternal Behaviour,* 65 ARCH. DIS. CHILD 1240 (1990).
63. S. D. Grosse et al., *Newborn Screening for Cystic Fibrosis: Evaluation of Benefits and Risks and Recommendations for State Newborn Screening Programs,* 53 MMWR RECOMM. REP. 1 (2004); M. H. Farrell & P. M. Farrell, *Newborn Screening for Cystic Fibrosis: Ensuring More Good than Harm,* 143 J. PEDIATR. 707 (2003).
64. L. N. Al-Jader et al., *Attitudes of Parents of Cystic Fibrosis Children towards Neonatal Screening and Antenatal Diagnosis,* 38 CLIN. GENET. 460 (1990).
65. A. Tluczek et al., *Psychosocial Risk Associated with Newborn Screening for Cystic Fibrosis: Parents' Experience while Awaiting the Sweat-Test Appointment,* 115 PEDIATRICS 1692 (2005).
66. Anonymous, *supra* note 51.
67. Tluczek, *supra* note 65.
68. Michie, *supra* note 58.
69. B. W. Forsyth et al., *The Child Vulnerability Scale: An Instrument to Measure Parental Perceptions of Child Vulnerability,* 21 J. PEDIATR. PSYCHOL. 89 (1996).
70. M. Green & A. J. Solnit, *Reactions to the Threatened Loss of a Child: A vulnerable Child Syndrome,* 34 PEDIATRICS 58 (1994).
71. *Id.*
72. M. B. Rothenberg & E. M. Sills, *Iatrogenesis: The PKU Anxiety Syndrome,* 7 JOURNAL OF THE AMERICAN ACADEMY OF CHILD PSYCHIATRY 689 (1968).
73. A. Bergman & S. Stamm, *The Morbidity of Cardiac Non-Disease in School Children,* 276 N. ENGL. J. MED 1008 (1967).
74. E. C. Allen et al., *Perception of Child Vulnerability among Mothers of Former Premature Infants,* 113 PEDIATRICS 267 (2004).
75. S. Bennett Johnson et al., *Maternal Anxiety Associated with Newborn Genetic Screening for Type 1 Diabetes,* 27 DIABETES CARE 392 (2004); K. K. Hood et al., *Depressive Symptoms in Mothers of Infants Identified as Genetically at Risk for Type 1 Diabetes,* 28 DIABETES CARE 1898 (2005).
76. M. S. Yu et al., *Impact on Maternal Parenting Stress of Receipt of Genetic Information Regarding Risk of Diabetes in Newborn Infants,* 86 AM. J. MED. GENET. 219 (1999).
77. Kerruish, *supra* note 10.
78. N. J. Kerruish et al., Maternal Psychosocial Reaction to Newborn Genetic Screening for Type 1 Diabetes, 2006 (unpublished).
79. Collins, *supra* note 30.
80. *Id.*

Newborn Screening Expansion: Massachusetts Research Models Encompass Public Health Service Responsibility

Anne Marie COMEAU

New England Newborn Screening Program, Massachusetts, USA

Massachusetts has been at the forefront of incorporating a formal research component into its newborn screening programme to advance our understanding of disease, its prevention and the related necessary services available to populations. This chapter advocates that research be recognized as an integral component of any newborn screening programme and that the infrastructure required within the programme to support such research be promoted as a provision of public health responsibility.

1. RESEARCH TO ADVANCE NEWBORN SCREENING

As early as the late 1990s, the New England Newborn Screening Program (NENSP) was authorized and directed by the Massachusetts Department of Health (MADPH) to increase the number of conditions in its newborn screening (NBS) panel dramatically, resulting in a list of conditions in Regulation[1] similar to that only recently recommended for a national uniform NBS panel.[2] Unlike the national uniform panel recommendation, the Massachusetts' deliberations[3] yielded a research component in addition to a mandatory component; >98% infants are screened for conditions included in the research component,[4] while essentially 100% infants are screened for conditions included in the mandatory list. The "mandatory" panel increased the number of conditions in the screen from 9 to 10^5 (more than any other state at the time of the expansion); the research or optional panel further increased that number by including another 20 conditions.[6] The optional panel is offered statewide for the duration of a simultaneous evaluation of feasibility and utility and is offered under a consent-based research protocol. The now seven year-old research protocol is in keeping with recent recommendations to "proceed with caution."[7]

We continue to learn from the research component. The approximate incidence of infants in the population identified as a result of Massachusetts' expanded newborn screening is 140/100 000 infants.[8] Of these, 25% are confirmed to have one of the conditions included in the statewide population-based research programme. The observed clinical outcomes for this subset of infants include a spectrum from "well" (relatively asymptomatic) through "death in later childhood" (preventable?) and "acute presentation with early death." As would be predicted in the implementation of a screening programme, the spectrum of the inherent disease associated with any particular condition identified by the screen is unveiled and the related clinical utility of the screen for the full range of the spectrum reveals new questions. These questions are especially pertinent when the candidate condition has a natural history and treatability that is known to providers by its later clinical presentation and a form of the disease is revealed by screening that has an unclear prognosis. Cystic fibrosis (CF) provides a good example.

The focus of the Massachusetts CF newborn screening programme has been to identify infants whose CF disease would benefit from early detection. In order to maximize sensitivity, the Massachusetts CF newborn screening programme incorporates a multimutation panel in its screening algorithm. Follow up to the screening has shown that a small percentage of infants with positive screens and two CFTR mutations have negative or borderline diagnostic tests.[9] Many of these infants were compound heterozygotes for a mutation associated with mild disease. Arguments about whether or not to include such presumably mild mutations in screening panels are confounded by observations of clinically concerning signs and symptoms in a small set of case studies,[10] which by definition have no controls for comparison. Thus, our understanding of the most common conditions—e.g. attributable genotype-phenotype relationships such as in our experience with CF screening—underscores the need to collect data on all conditions in the panel.

Since the 1999 implementation of Massachusetts expanded newborn screening, and in great part since the 2005 American College of Medical Genetics (ACMG) recommendations for a uniform panel,[11] 39 other states have increased the number of conditions included in their newborn screening programmes significantly;[12] all 39 currently screen or plan to implement expanded screening for conditions that are included in the Massachusetts research component as part of routine (non research) newborn screening panels in those states. Such expansion by multiple states does begin to address an issue of equity in U.S. public health practices. More infants will be screened for a similar set of disorders regardless of state borders. However, such expansion is dependent on a national call for standardization and does not necessarily ensure that the evidence base we need for improvement of services will be established. Other issues of equity have yet to be addressed: how do we ensure that the conditions for which the public is most likely to benefit from screening are the conditions screened vs. conditions that happen to be advocated for by well-funded professional and lay organizations? To ensure this requires an evidence base. Such an evidence base requires an infrastructure within each newborn screening programme that supports research on equal footing with basic newborn screening programme services.

2. INFRASTRUCTURE FOR IMPLEMENTATION OF SCREENING FOR NEWLY ELIGIBLE EMERGING CONDITIONS

The Secretary's Advisory Committee on Heritable Disorders and Genetic Diseases in Newborns and Children (ACHDGDNC) is deliberating the process by which new conditions are added to the uniform panel.[13] The nomination form that is in development would require submission of evidence about treatment, the screening test, the diagnostic test and clinical outcomes. This is excellent progress toward standardizations of the criteria by which conditions are added to population-based screens. It is reasonable that common criteria derived from a solid evidence base be used. Of course, in the absence of population-based screening, no data are generated for the evidence base. In the absence of sufficient evidence then, and in order to access sufficient evidence, population-based research will be required. The Massachusetts model for consent-based population research provides an infrastructure by which such evidence can be collected prior to the decision that a condition qualifies for population-based screening. Such a model provides an infrastructure for the implementation of Stage I and II Research called for by Botkin.[14] Rather than preventing implementation of screening for a disorder, such research is likely to facilitate population-based access to cutting edge screening algorithms: when there is a promising treatment for a condition that has an assay that appears to be applicable to newborns, a pilot programme can be implemented to test the clinical validity of the screen and efficacy of the treatment. Furthermore, if the Massachusetts model for consent-based population research were to be incorporated in other states, evidence for a variety of conditions might be collected in a more timely manner; though the research component would not necessarily ensure equity across all states, it would ensure a dynamic opportunity for evaluation of multiple screening protocols in multiple states.

3.   INFRASTRUCTURE FOR FOLLOW UP AND DATA MANAGEMENT (WHAT IS THE CLINICAL UTILITY OF THE SCREEN?)

The essentials of newborn screening programmes provide a centralized system for ensuring that infants are screened, identifying affected infants, and tracking affected infants identified by the screen to diagnosis and treatment.[15] The infrastructure for coordinating these services is in place in all newborn screening programmes. In addition, some programmes include long-term follow up activities for some conditions included in the screen. Long-term follow up activities include evaluation of compliance with treatment, evaluation of clinical outcomes associated with treatment(s), and evaluation of effects of adulthood and lifestyle choices not typically associated with pediatric evaluation (e.g. pregnancy).[16] Infrastructure for comprehensive long-term follow up activities is not well developed in most newborn screening programmes and is dependent on sophisticated information technology systems typically not supported by funds dedicated to routine newborn screening. The task is challenging and will require careful policy development matched with technology. Simple projections suggest that that in 20 years' time a programme that currently identifies 160 new cases each year would carry the responsibility for re-finding and following up on a minimum of 3200 infants each year in addition to finding the new 160 infants (this assumes that no additional conditions were included in the interim). It seems reasonable policy to consider limitations on the long-term tracking of infants with conditions and treatments that are well understood and to focus maximum effort on the long-term tracking of infants with conditions added to panels or with familiar conditions with new treatments.

Independent of the issues of local infrastructure regarding information technology and financial support for labour-intensive efforts are the very real issues of the need to establish case definitions supported by firm laboratory and clinical parameters. Raw data from these parameters will have to be collected and maintained in order to allow retrospective re-evaluation of the cohort meeting the case definition. In order to ensure multi-site cohort comparison, multidisciplinary partners representing laboratories, datasystems and clinical perspectives will have to set up minimal data criteria as has been done for a variety of federally funded epidemiologic studies of disease.

4.   INFRASTRUCTURE FOR COORDINATION OF CLINICAL TRIALS
(WHERE ARE THE CASES, WHAT IS THEIR STATUS, HOW TO ENROLL IN
CLINICAL TRIAL)

Newborn screening programmes have traditionally only offered screening services for conditions meeting Wilson and Jungner criteria.[17] When such criteria are not met, or when it is unclear that early detection provides direct benefit to the infant affected with the condition, there should be acknowledgement of this lack of knowledge to guardians of the infants. When research is the justification for the screening, such research must be conducted pursuant to the Federal Common Rule 45 CFR 46[18] governing research on human subjects. As discussed above, the research might focus on the feasibility of implementing a screening programme for a newly eligible condition. Such research is clearly an integral part of the newborn screening programme. In addition, the research might actually focus on the generation of a cohort of infants who would be available for a clinical trial. It is unprecedented that a population-based and state-authorized programme would provide a service as a means to generate a cohort for a clinical trial. Such an unprecedented impetus may very well be necessary to gain the knowledge that we need to advance the science of rare pediatric conditions.

Clinical trials for prevention of early onset pediatric conditions are faced with particular challenges in recruiting individuals prior to the onset of sequelae and irreversible damage. Infants and children whose condition is particularly rare may be called upon repeatedly for participation in multiple clinical trials or clinical trials for their condition may be hindered due to lack of a reasonable cohort. Newborn screening programmes that identify and follow infants and children with conditions for which clinical trials are planned have the potential to offer a centralized resource for coordination of clinical trials. To do so, any newborn screening programme that would provide a screening service as a means to generate a cohort for a clinical trial must be open and true to the public it serves, with appropriate notice of the purpose of the screen. Such a screening service should be treated as a research protocol in and of itself. Two types of these services can be envisioned: those for which a specific endpoint or specific clinical trial are planned and those for which no specific clinical trial is planned, but entry into a registry is a byproduct of the positive screen.

Enhancement of the existing infrastructure within the newborn screening programme would negate the need to duplicate the effort of setting up separate registries for each disorder type. Unlike multiple registries existing at or generated by a variety of specialty care centres, the newborn screening datasystem has well-protected information on all infants with newborn screening disorders. One could envision a protocol by which parents agree to be re-contacted by the newborn screening programme if and when a clinical trial is beginning. When a clinical trial meeting national standards is recruiting participants, the clinical trial's research coordinator might send notice to state newborn screening programmes who would then in turn send a notice of the trial to the relevant individuals, inviting them to obtain more information about the trial from the trial's research coordinator. An advantage of this system is that it would streamline evaluation of eligible participants by the use of the long-term follow up fields residing in newborn screening datasystems, which are discussed above, for the evaluation of current status and of whether or not the individual meets trial criteria. The system would also provide notice of the trial to all eligible patients, regardless of primary and specialty provider or geographic location. Finally, it would provide a notice of the trial without perceived pressure from a treating physician/centre.

5.   INFRASTRUCTURE FOR EPIDEMIOLOGIC EVALUATIONS

The first major research initiative within an existing newborn screening programme focused on the epidemiologic evaluation of the number of HIV-infected childbearing women as a surrogate marker for monitoring trends in the HIV epidemic.[19] The serosurvey, later adopted nationally, provides a model for deidentified use of residual dried blood spots. By the time the national study was discontinued in 1995, more than 12 million births nationwide had been sampled, indicating geographic pockets of high seroprevalence and geographic regions of low seroprevalence that public health authorities used for planning. There was no loss of confidentiality. Since that time, other investigations have made use of residual dried blood spots to estimate the frequency of a condition, but no subsequent study approached the monumental effort that injected technology development and sophisticated laboratory personnel into public health laboratories.

6.   OVERARCHING INFRASTRUCTURE

Newborn screening programmes and systems provide a strong foundation on which to build the research infrastructure to expand our understanding of pediatric disease and disease prevention. Investigators with true operational responsibilities for population-based newborn screening programmes as well as clinical investigators must be included in the planning, implementation and analyses of such research. Policies to ensure public trust, collegial relationships, interdisciplinary collaborations, data sharing and public access will have to be developed; policies that parallel the GAIN policies[20] should be explored.

REFERENCES

1.   105 MASS. CODE REGS. 270.000.
2.   American College of Medical Genetics, Newborn Screening: Toward a Uniform Screening Panel and System. Report for Public Comment (2005), http://www.mchb.hrsa.gov/screening/.
3.   K. Atkinson et al., *A Public Health Response to Emerging Technology: Expansion of the Massachusetts Newborn Screening Program*, 116 PUBLIC HEALTH REP. 122 (2001).
4.   A. M. Comeau & D. E. Levin, *Population-Based re Population-Based Research within a Public Health Service: Two Models for Common Rule Compliance in the Massachusetts Newborn Screening Program* (forthcoming 2006).
5.   PKU, maple syrup urine disease, homocystinuria, galactosemia, congenital hypothyroidism, congenital toxoplasmosis, congenital adrenal hyperplasia, sickle cell disease, biotinidase deficiency, and medium-chain acyl-Co-A dehydrogenase (MCAD) deficiency.
6.   Cystic fibrosis (CF), Tyr I (Tyrosinemia I), Tyr II (Tyrosinemia II), ß-MC (ß-Methyl Crotonyl Carboxylase), HMG (HMG Lyase Deficiency), ASL (Argininosuccinic Aciduria), IVA (Isovaleric Acidemia), HHH (HHH Syndrome), GA I (Glutaric Acidemia I), GA II (Glutaric Acidemia II), ASS (Citrullinemia), MMA (Methylmalonic Aciduria), PPA (Propionic Acidemia), CPT II (CPT Deficiency), LCHAD (long-chain hydroxy-CoA dehydrogenase deficiency), VLCAD (very-long-chain acyl-CoA dehydrogenase deficiency), SCAD (short-chain acyl-CoA dehydrogenase deficiency), LCAD (long-chain acyl-CoA dehydrogenase deficiency), ß-KT (ß-Ketothiolase Deficiency), Argininemia. Note that the list approved by metabolic specialists includes LCAD and that recent deliberations have indicated these cases to be VLCAD. Note that the list of conditions does not include those conditions detected as a result of screening for the MADPH panel.

7. J. R. Botkin et al., *Newborn Screening Technology: Proceed with Caution,* 117 PEDIATRICS 1793 (2006).
8. A. M. Comeau et al., *Integration of New Genetic Diseases into Statewide Newborn Screening: New England Experience,* 125C AM. J. MED. GENET. 35 (2004).
9. R. B. Parad & A. M. Comeau, *Diagnostic Dilemmas Resulting from the Immunoreactive Trypsinogen/DNA Cystic Fibrosis Newborn Screening Algorithm,* 147 J. PEDIATRICS S78 (2005).
10. B. P. O'Sullivan et al., *Early Pulmonary Manifestation of Cystic Fibrosis in Children with the ΔF508/R117H-7T Genotype,* PEDIATRICS (forthcoming 2006).
11. American College of Medical Genetics, *supra* note 2.
12. National Newborn Screening and Genetics Resource Center (NNSGRC), http://genes-r-us.uthscsa.edu/; March of Dimes, http://www.marchofdimes.com/ peristats/pdfdocs/nbs2006.pdf.
13. Secretary's Advisory Committee on Heritable Disorders and Genetic Diseases in Newborns and Children, Summary of Seventh Meeting, Feb. 13-14, 2006, http://mchb.hrsa.gov/programs/genetics/committee/7thminutes.htm#6.
14. J. R. Botkin, *Research for Newborn Screening: Developing a National Framework,* 116 PEDIATRICS 862 (2005).
15. Clinical and Laboratory Standards Institute, *Newborn Screening Follow Up: Approved Guideline,* CLSI document I/LA27-A (2006).
16. P. J. Lee et al., *Maternal Phenylketonuria: Report from the United Kingdom Registry 1978-1997,* 90 ARCH. DIS. CHILD. 114 (2005); J. E. Haddow et al., *Maternal Thyroid Deficiency During Pregnancy and Subsequent Psychological Development in the Child,* 341 N. ENGL. J. MED. 549 (1999).
17. J. M. G. Wilson & G. Jungner, Principles and Practice of Screening for Diseases (1968).
18. 45 C.F.R. 46.
19. R. Hoff et al., *Seroprevalence of Human Immunodeficiency Virus among Childbearing Women. Estimation by Testing Samples of Blood from Newborns,* 318 N. ENGL. J. MED. 525 (1988).
20. Genetic Association Information Network (GAIN), http://www.fnih.org/GAIN/ policies.shtml.

Carriers Detected by Neonatal Screening: A Clinical Geneticist's View

Helena KÄÄRIÄINEN

Department of Medical Genetics, University of Turku, Turku, Finland

Ilona AUTTI-RÄMÖ

Finnish Office for Health Technology Assessment, National Research and Development Centre for Welfare and Health (STAKES), Helsinki, Finland

1. INTRODUCTION

The aim of neonatal screening is usually to detect infants with a severe but treatable diseases early enough to prevent serious outcomes, like irreversible organ damage, mental handicap or death. In some exceptional cases, neonatal screening programmes (usually pilot programmes) have been instituted with the aim of diagnosing severe, hereditary conditions in a family early enough for family planning and, also, to give parents time to adapt to the situation.

In the case of neonatal screening of metabolic disorders, the aim is clearly to prevent morbidity due to disease. If the aim were to find carriers and thus prevent affected cases from being born, a more efficient way of screening would be carrier screening of young adults before their first pregnancy. The different options for screening or diagnosing such diseases are schematically depicted in Figure 1.

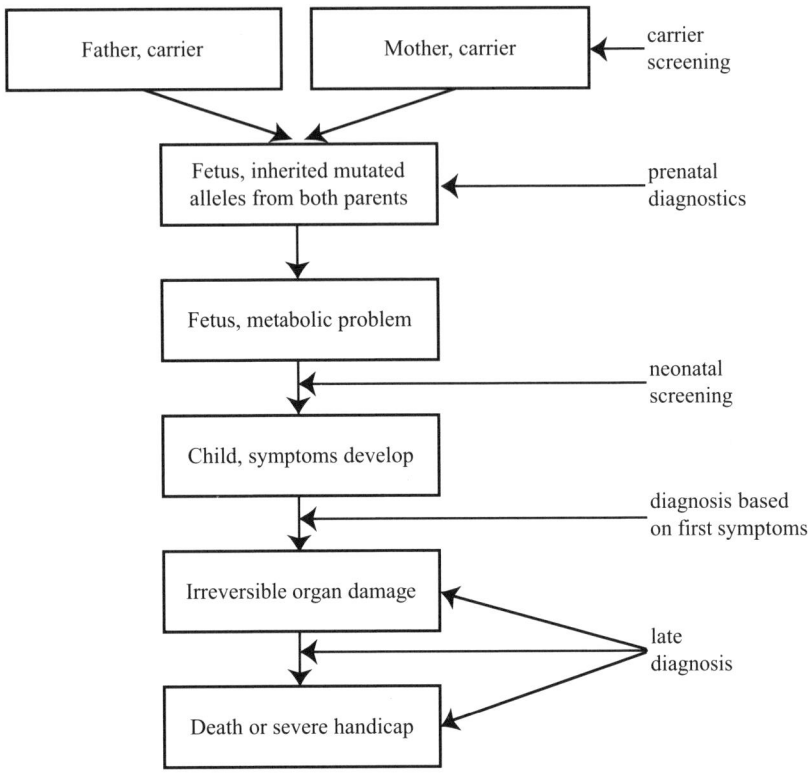

Figure 1. In the case of autosomal recessive metabolic diseases, both parents are asymptomatic carriers and their offspring have a 25% probability of inheriting the mutated gene from both parents and developing a symptomatic disease. Carrier screening and prenatal diagnosis would offer the possibility of preventing the affected children from being born. Neonatal screening and early treatment, when available, prevent the severe consequences of the disease. Some of the metabolic diseases can still be treated successfully when diagnosis is made soon after the first symptoms. In some other diseases, irreversible damage has already occurred at that time and severe handicap or death will follow because the possible treatments have been started too late.

The aim of neonatal screening of treatable metabolic diseases is thus ethically straightforward, and it is generally accepted by parents, healthcare personnel and politicians. The difficult ethical questions related to carrier screening programmes (with the aim of offering the option of terminating pregnancy) are not directly raised in neonatal screening programmes. However, many of the families want to choose prenatal diagnostics in future pregnancies.

In spite of this acceptable goal, neonatal screening programmes also create serious problems. One is, of course, the problem of costs: even though preventing morbidity (or death) saves some costs in the long run, the testing procedure, with training of personnel, information for and counselling of the couples, optimal logistics, best possible treatment of the conditions, etc, creates costs, and the cost-benefit calculations are complicated by the many, diverse benefits and harms which are hard to measure or compare with one another.[1] In addition, offering a screening programme to detect serious diseases during the neonatal period will inevitably cause some worry and anxiety, the long term consequences of which are impossible to measure. Finally, in addition to real false positives, healthy carrier siblings may be inadvertently detected. The problem of detecting carrier newborns who will never get symptoms of the disease is the topic of the present paper.

2.   NEONATAL SCREENING IN FINLAND

In Finland, the only metabolic disease currently being screened for in newborns is hypothyreosis; this screening is performed from cord blood. Most Western countries offer much wider screening programmes for newborns. In particular, phenylketonuria (PKU) is screened in neonatal period in practically all Western countries.

In Finland, PKU is known to be rare. Cases have been systematically searched for by screening mentally retarded individuals in institutions and healthy newborns. These studies, performed some 30 years ago, resulted in an approximation of the incidence of PKU to be 1/100 000 – 1/200 000 newborns. Because of the rarity of PKU, it is not screened for in newborns in Finland.[2]

Usually, the extended screening programmes are built onto the existing PKU screening. PKU screening is very efficient in preventing mental handicap in the affected children and thus considered cost-efficient in countries where the incidence of PKU is relatively high. As the samples are taken and sent to a screening laboratory anyway, adding other diseases with less well-proven beneficial effects of early diagnosis does not add to the costs very much. Thus, in many countries where PKU-screening is offered, screening for other (metabolic) diseases has been added to the programme.

Recently, immigration into Finland from various countries has increased, creating a situation in which newborn screening for PKU has to be offered for non-Finnish couples anyway. At the same time, the new possibilities of screening for several metabolic diseases simultaneously using tandem mass spectrometry prompted a health technology assessment project on the effect and costs of expanded newborn screening.[3] The evaluation found that, with a selection of diseases consisting of PKU, MCAD, LCHAD, glutaricaciduria, congenital adrenal hyperplasia, the costs per quality-adjusted life year (QALY) gained would be a maximum of €25 500 and, after discussion, a decision was made not to start expanded neonatal screening at this point. In addition to real monetary costs, untoward side-effects of the possible screening programme were evaluated, one of those being inadvertently detecting carrier newborns.

3.   CARRIERS OF AUTOSOMAL RECESSIVE DISEASES IN NEONATAL SCREENING

In case of autosomal recessive inheritance, a child will be affected only if he or she gets a mutated gene from both parents. Such a carrier couple has each time a higher chance (50%) of having a carrier and a lower chance (25%) of having an affected baby (Figure 2).

Autosomal recessive inheritance

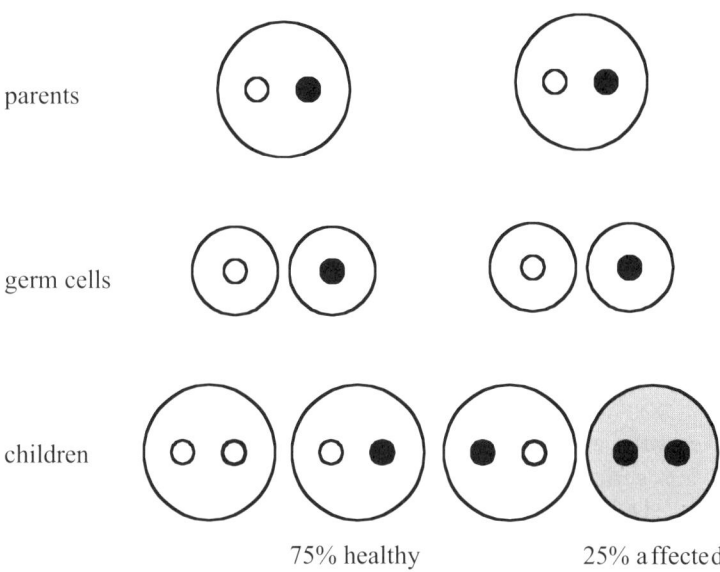

Figure 2. Carrier parents have one normal allele (white dot) and one mutated allele (black dot). In autosomal recessive inheritance, one mutated allele does not lead to any symptoms but instead to being a carrier. If the child happens to inherit a mutated allele from both parents, he or she will be affected.

Furthermore, most carriers never marry another carrier and are never at risk of having affected children. Instead, half of their children are carriers. For instance, if the incidence of carriers in a population is 1/10 (a very high incidence), then only 1/100 of couples are carrier couples and, in their each pregnancy, the likelihood of an affected child is 25%. This high number of asymptomatic carriers in the population of all recessive diseases, in relation to the number of those affected, leads to the fact that any newborn screening programme will find many more carriers than affected individuals if the screening method is such that carriers will also be detected.

By definition, autosomal recessive inheritance means that the heterozygote carrier's phenotype cannot be differentiated from that of the normal homozygote. Often, however, some special methods reveal slight abnormalities, like half amounts of the protein concerned in the carriers when compared to normal individuals. Sometimes, mild phenotypic manifestations have also been detected.[4] If carriers have symptoms of disease, the philosophical question arises whether the mode of inheritance can be called autosomal recessive any more.

On the contrary, there are several examples of situations in which heterozygous carriers have a "heterozygote advantage;" this is the most common explanation for the fact that some recessive mutations have become so prevalent in the population. The best known example is the protective effect of being a β-thalassemia carrier against malaria.[5]

Recently, at least one example has shown that being a carrier for a rare autosomal recessive disease may be a risk factor for another disease. This is the case in Fanconi anemia, where one of the genes, FANCD1, is identical to BRCA2, and thus being a heterozygous carrier of FANCD1 mutations means clearly elevated risk of breast and some other cancers.[6] This implies that, in the future, some other situations of being "symptomless carriers" may turn out to be associated with serious health consequences for the individual.

4.   POSITIVE AND NEGATIVE CONSEQUENCES OF FINDING A CARRIER IN NEWBORN SCREENING

Discovering a newborn is a carrier may have both positive and negative consequences. The shock of being told that something— anything—was found in the screening may create a lot of anxiety in the newborn period, when parents can be exceptionally sensitive and vulnerable. This anxiety should usually be alleviated with prompt and comprehensive counselling, while some parents may remain anxious even after the best possible information and counselling. In addition, there could be situations where after finding one mutation and not another, the situation does not with certainty mean that the child is "only" a carrier. On the contrary, there may be another mutation that has not been detected. Thus, it is not always possible to assure the parents that the child will definitely not be affected.

Except for the immediate anxiety, it could be argued that parents and the family benefit from the detection of carrier status in their child. This information tells them immediately that at least one of the parents must be a carrier as well, and they may even be a carrier couple. In the latter case, there could be an older child in the family who might benefit from diagnosis of the disease concerned. For instance, LCHAD could be still asymptomatic in an older sibling. Also, the parents could choose to terminate a future pregnancy. Detecting carriership in a parent may also lead to cascade screening of more distant relatives, if the family wishes.

One could argue that finding carriers is also a good and useful thing for the newborn because, while growing up, he or she will know the situation and will be able to request his or her partner to be mutation-screened as well. Thus, carrier couples could be detected in good time and birth of affected children could be prevented.

It has, however, been shown that finding mutation carriers in childhood does not necessarily lead to correct information for the child at an optimal time as a young adult or correct understanding of the information.[7] The parents may misunderstand the situation and convey the message incorrectly, choose not to tell about the carriership at all or even forget about it. In cases where the parents have told about the carriership to the child, many of them have felt this task very difficult and demanding.[8]

Detecting carriership in a newborn definitely takes away the child's right not to know and also the right to privacy: as a young adult, the child might have chosen not to disclose the result of a carrier test, for instance, in order to safeguard privacy in making decisions about possible prenatal testing in future pregnancies.[9] In addition, healthcare may develop in unforeseen directions, which may mean that a newborn detected as a carrier today will, in the future, be obliged by healthcare funding systems or others to participate in prenatal diagnostics even though this would be against personal ethical principles. Furthermore, the fact that one knows about carriership for a recessive disease at young age may, in an extreme situation, cause feelings of stigmatization, disturb one's self-esteem and form an obstacle in starting relationships and committing to them.[10]

5.   TO SCREEN OR NOT TO SCREEN?

When screening programmes are offered to the general public, it usually means that most individuals get a normal result. Looking at the situation retrospectively, those individuals never benefited from the screening programme. The ones who were found to have the conditions screened for are the ones who get the benefit. Those who got results that are difficult to interpret or who were given some information that was initially not the goal of the program (like finding carriers in neonatal screening) are the ones who actually receive no benefit but instead some harm.

This happens to some extent in most screening programmes. In a way, the well-being of these individuals is partially sacrificed in order to get the important benefit to the others. As stated by the Tavistock group, the complexity and cost of a healthcare delivery system may set up tension between what is good for society as a whole and what is best for individuals.[11]

The four ethical principles of the healthcare—beneficence, non maleficence, respect for autonomy and justice—are all important, but the hierarchy between them is not self-evident and varies in different healthcare situations.

When looking at the neonatal screening programmes from the detected carrier's perspective, the possible options would be one of the following:

1) No screening is performed and carrier status will remain undetected, which will have both positive and negative consequences, as discussed above.

2) Screening is performed, carriers are detected, and this leads to the positive consequence of detecting a carriers/carrier families and being able to get genetic counselling, as well as the negative consequences of carrier testing in childhood, including testing before being able to give consent to it.

6.   A CLINICAL GENETICIST'S VIEW

The problem of detecting carriers in newborn screening programmes is, in the end, a problem of pre-test and post-test information and genetic counselling. Optimally, the possibility of detecting carriers should be discussed with the parents before the screening test because it might affect their willingness to consent. In reality, however, comprehensive pre-test information is often not given in population screening programmes, and false positive results or detecting carriers may come as an unexpected surprise to the parents.

An option to solve this problem would be to avoid detecting carriers by choosing a screening method, if possible, that does not detect carriers. Such methods are available for some diseases (for instance, cystic fibrosis), but there are also other differences between the available screening methods, like earlier or later availability of results, which complicate the choice. Another possibility would be to not disclose carrier status. This would abolish the problem of inadvertent newborn carrier testing but also prevent families from the possible benefits of detecting a carrier baby. If such an option is chosen, the situation should, naturally, be discussed with the couple at the time of the consent process. The third possibility is to disclose the carrier status but, simultaneously, try to find the best ways of doing that.

According to a recent review, there are no controlled trials about disclosing carrier status in newborn screening programmes.[12] There appears to be more opinions than real data concerning the consequences of newborn carrier detection and thus a need to investigate the issue and to develop and evaluate counselling and support of the couples in these situations.

When planning, performing and evaluating (newborn) screening programmes, special emphasis should be given to the availability of comprehensive pre-test information for those who want it and ample resources for immediate support and genetic counselling in case of results suggesting the baby to be either affected or a carrier.

REFERENCES

1. N. J. Kerruish & S. P. Robertson, *Newborn Screening: New Developments, New Dilemmas,* 31 J. MED. ETHICS 393 (2005).
2. J. Palo, *Prevalence of Phenylketonuria and Some Other Metabolic Disorders Among Mentally Retarded Patients in Finland,* 43 ACTA NEUROL. SCAND. 573 (1967); J. K. Visakorpi & O. V. Renkonen, *The Incidence of PKU in Finland,* 60 ACTA PAEDIATR. SCAND. 666 (1971); O. Simell, *Phenylketonuria in Finland–A Population Genetics Rarity?* 71 LAKARTIDNINGEN 1170 (1974).
3. I. Autti-Rämö et al., *Expanding Screening for Rare Metabolic Disease in the Newborn: An Analysis of Costs, Effect and Ethical Consequences for Decision Making in Finland,* 94 ACTA PAEDIATR. 1126 (2005).
4. M. Arvio et al., *Dysmorphic Facial Features in Aspartylglucosaminuria Patients and Carriers,* 13 CLIN. DYSMORPH. 11 (2004).
5. J. Flint et al., *Why Are Some Genetic Diseases Common? Distinguishing Selection from Other Processes by Molecular Analysis of Globin Gene Variants,* 91 HUM. GENET. 91 (1993).
6. D. Papadopoulos & E. Moustacchi, *L'anémie de Fanconi : gènes et fonction(s) revisités,* 21 MED. SCI. (PARIS) 730 (2005).
7. O. Järvinen et al., *Carrier Testing of Children for Two X-Linked Diseases: A Retrospective Study of Comprehension of the Test Result and Social and Psychological Significance of the Testing,* 106 PEDIATRICS 1460 (2000); O. Järvinen et al., *Carrier Testing of Children for Two X-Linked Diseases in a Family-Based Setting: A Retrospective Long-Term Psychosocial Evaluation,* 36 J. MED. GENET. 615 (1999); J. Mitchell et al., *What Young People Think and Do when the Option for Cystic Fibrosis Carrier Testing is Available,* 30 J. MED. GENET. 538 (1993); A. Jolly et al., *Testing Children to Identify Carriers of Balanced Chromosomal Translocations: A Retrospective, Qualitative Psychosocial Study,* 33 J. MED. GENET. (supplement) abstract 6.018 (1996).
8. Järvinen (2000), *supra* note 7.
9. Working Party of the Clinical Genetics Society, *The Genetic Testing of Children,* 31 J. MED. GENET. 785 (1994).

10. Jolly, *supra* note 7.
11. Tavistock Group, *Shared Ethical Principles for Everybody in Healthcare,* 318 B. M. J. 248 (1999).
12. S. Oliver et al., *Disclosing to Parents Newborn Carrier Status Identified by Routine Blood Spot Screening,* COCHRANE DATABASE SYS. REV. CD003859 (2004).

Systematic Neonatal Screening and Carrier Detection: Lessons from Sickle Cell Disease and Cystic Fibrosis Screening in France

Jean-Louis DHONDT

French Association for the Detection and Prevention of Metabolic Diseases and Handicaps in Children
Laboratoire de Biochimie, Hôpital St Philibert, France

In general, newborn screening is recommended for disorders which fulfill the following criteria: (1) there is considered to be a direct benefit to the newborn from early diagnosis, (2) the benefit is reasonably balanced against financial and other costs (at both the individual and population levels), (3) there is a reliable test suitable for neonatal screening, and (4) there is a satisfactory system in operation to deal with diagnostic testing, treatment and follow-up of identified babies.

This is best exemplified by a disorder like phenylketonuria (PKU), in which early detection through screening and treatment has been very effective in preventing neurological abnormalities. In many countries, screening has included the search for various other conditions, but some methods inadvertently identify newborn infants who, although not affected by the condition, carry a gene for it. The discovery of carrier status can lead to testing parents and family members. Such a possibility is an important change in the concept of neonatal screening, at least as initially defined. Consequently, establishing guidance for disclosing or not disclosing carrier status is essential but must involve professionals (screeners and geneticists), the general population and public health decision-makers. The purpose of this chapter is to describe the different aspects and problems of carrier identification in neonatal screening programmes based on the experience of the French programme.

1.   THE FRENCH SCREENING PROGRAMME

By 1965, programmes for newborn screening of phenylketonuria (PKU) had been developed locally in France. The results of these programmes, the promising development of tests for the screening of other diseases and the growing perception of the need for a national policy for screening led to the creation, in 1975, of the "French Association for the Detection and Prevention of Metabolic Diseases and Handicaps in Children" (AFDPHE),[1] which is a private association that assumed responsibility for public health programmes under the tutelage of the Ministry of Health. An agreement with the social security agency (CNAMTS, the National Fund for the Medical Insurance of Salaried Workers) defines the preventive programme to be executed and specifies the financial support.

The programme started with screening for PKU, then added screening for congenital hypothyroidism (CH, 1978) and congenital adrenal hyperplasia (CAH, 1981), three conditions which fulfilled the "prerequisite" for systematic screening.

Systematic screening for haemoglobinopathies began in 1985 in high-risk areas overseas (West Indies and French Guiana). In 1990, an evaluative programme was developed to examine the possibility of systematic screening for sickle cell disease (SCD) in metropolitan France, and, in 2000, a targeted programme was implemented because of the great variation in distribution of at-risk populations (20% in Paris compared to 3% in Brittany).

From 1989 to 1990, a pilot programme[2] was conducted in order to evaluate the feasibility of cystic fibrosis (CF) screening. Based on 513 000 tests, the conclusion was not in favour of the screening using the strategy available at that time (measurement of immunoreactive trypsinogen (IRT) in neonatal blood at day 3 and control at day 21) because the recall rate was higher than anticipated (1.9% of babies required analysis of a second blood sample) and not acceptable due to the possible adverse psychological effects on so many families of infants with false positive results. Since then, the IRT/DNA testing algorithm has improved the specificity of CF screening. In 2000, AFDPHE was mandated by its regulatory agencies to organize systematic screening for CF in France; this programme started in 2002, after a reorganization of the CF centres.[3]

Table 1 below, presents the main characteristics of the French neonatal screening programme.

2.   GENETIC TESTING AND NEONATAL SCREENING

The screening of PKU, CH and CAH are based on biochemical phenotype recognition (respectively, elevated phenylalanine, TSH, 17 hydroxy-progesterone). CH has an unknown genetic background (if one exists), while PKU and CAH are both autosomal recessive diseases for which the most frequent mutations are known. Although DNA assay of dried blood spots (DBS) is possible, the inclusion of DNA testing in PKU screening has never been considered on a systematic basis, mainly because of the great number of possible mutations (463 reported in 2003). For CAH, the search for the most frequent mutations in the *CYP21* gene has been proposed as a confirmatory test, but not really performed on a routine basis.

Screening for SCD was the first to give a direct insight in the genetic status of the tested baby, since the characterization of haemoglobin variants (isoelectric focusing, HPLC) can identify carrier status.

For the screening of CF, most screening programmes in the world experienced the limits of relying only on the IRT assay, and adopted the inclusion of a DNA test (initially the search of delF508 then multiple-CFTR-mutation testing), thus providing a model "to investigate the implications of applying multiple-mutation DNA testing in screening for any disorder in a pediatric population-based setting, where detection of affected infants is desired and identification of unaffected carriers is not."[4] Several mutations depend on ethnic background.[5] For example, to cover a maximum of potential mutations, France added the mutation Y122X to the panel, which is frequent in the population of La Reunion Island (40% of identified alleles).

Table 2 below, summarizes the difference between SCD and CF. It is essential to specify that the methods used for screening affected infants have lower performance in identifying carriers. In SCD screening, almost 100% of A/S, A/C heterozygotes are recognized (4% of tested newborns), but the method does not identify carriers for β thalassemia. In CF screening, only infants with high IRT have a search for mutations (with a panel of 30 mutations which cover 86% of mutations); consequently, only 1% to 2% of CF heterozygotes are identified (1419 carriers identified out of 2,717,992 screened infants, 2002-2005).

3.   OFFICIAL REGULATIONS AND FRENCH "BIOETHICS" LAWS

Several laws and decrees have regulated the prescription of genome tests and the communication of their results.

In 1978, Law 78-17[6] was enacted concerning the processing of nominative data, data banks and individual liberty. This law also set out that "persons whose health data is likely to be used must be informed of the purpose of such usage."

Enacted in 1988, Law 88-1138[7] defined the role of the "Advisory Committee for the Protection of the Individual" which must be consulted before undertaking any biochemical research. This law also sets out a specific definition for free, informed and written consent.

In 1994, two laws (94-653[8] and 94-654[9]) introduced the notion of the "study of genetic characteristics of an individual" in the context of the respect for the "human body."

The Law of 1996[10] and a decree published in 2000[11] describe the agreements required for the prescription of any DNA testing, as well as for the communication of results. The main points are the following:

1) DNA analysis can be performed only if the result implies medical intervention with a direct benefit for the patient (a genetic study can be prescribed only for symptomatic subjects or at-risk subjects (family context));

2) Informed consent must be obtained from the patient or from the patient's parents or legal guardians during a medical consultation aimed to inform the patient or the parents/guardian about the disease being tested for, the meaning of the genetic testing, and medical consequences if the disease is confirmed. A document which attests that the information has been given, signed by both the patient (or parents/guardian) and the physician, is sent to the laboratory which will perform the test, and a copy must be kept in the medical record;

3) Laboratories have to be authorized by the Ministry of Health;

4) The results are sent only to the physician who prescribed the test; and

5) The results can be announced only during a special consultation.

In response to criticisms and lack of specific legal provisions, a new law was enacted in 2004.[12] For the first time, "screening for handicaps" is considered, but the articles of the decree which will specify the practical aspects are still in a "draft" version circulating for comments from experts.

It is remarkable that in none of these texts is the communication of carrier status considered.

4.   INFORMATION AND WRITTEN CONSENT

In most parts of the world, classical newborn screening for routine/treatable disorders has been and is being carried out without explicit parental consent. In newborn screening programmes, consent is presumed and justified on the basis that when a disease is treatable, a newborn has a right to be screened and to be treated. Most newborn screening programmes are part of mandated paediatric norms and are considered part of routine care. In France, based on these principles, the screening programme for PKU, CH, CAH, SCD does not require written consent and has not since the beginning of the programme ("tacit consent")[13] (in the case of SCD screening, the method in use is not considered a "genetic test" as defined by law). When parents refuse the tests, it is recommended that a written dissent be kept in the medical record and recorded on the filter paper card (to inform the screening centre). The principle of systematic information on the screening programme has been adopted for many years, by means of a leaflet given with oral comments at the time of the collection of the DBS.

However, the inclusion of DNA-based testing techniques in the CF screening algorithm required procedures to be brought into line with the French laws on bioethics. Consent must be obtained. For practical reasons, it was decided that written consent be collected systematically for all neonates at birth by having the parents sign directly on the sampling card. The text of the consent ("After being informed, we the undersigned mother, father of the child (name:...) hereby authorize □ do not authorize □ the physicians responsible for neonatal screening to perform a genetic test for cystic fibrosis if necessary") has been established according to the recommendations of the AFDPHE ethics committee. To avoid the "misuse" of cards for any other DNA testing, it has been decided to specify on the form that the consent of parents was restricted to the CFTR gene testing in case of a positive IRT assay.

New educational materials were designed for parents; these approach screening in a general manner with equal emphasis on the diseases covered and present the screening tests as a routine matter to minimize anxiety. The only document related to the possibility of identifying carriers has been designed for the screening of SCD and is given to the parents who belong to at-risk groups.

A practical guide for professionals is also distributed in maternity wards, in order to prepare health professionals for participating in the informing process.

5.   WHAT TO DO AFTER CARRIER IDENTIFICATION

Withholding such information is still debated, except in countries where it is clearly against legislative guidelines. Many aspects (both pros and cons) have been identified:

5.1   *Ethical Aspects*

Respect for individual rights: "The right to know *vs* the right to decide." The new Law on bioethics of 2004[14] states that when the diagnosis of a severe genetic disease is made, the physician must inform the patient about the risk of not disclosing the result to other family members. But it is his or her right to disclose or not disclose his or her genetic status!

Avoiding parental anxiety: Some health professionals consider the identification of healthy carrier infants as undesirable because of the potential for unjustified anxiety about the health of the newborn and disruption to the mother/baby relationship. Some studies concluded that carrier identification was not always perceived by parents to be problematic;[15] however, no controlled trials about disclosing carrier status were found in a database search by Oliver et al.[16]

The risk of discrimination or stigmatization: Information about carrier status can raise fears of stigmatization (misuse by insurers, employers, etc)—fears which are evident in many "ethnic minorities" (labeling a person or family as having "undesirable" characteristics)—although discrimination is illegal in France (cf.

the "politically correct" proposal to define such ethnic groups as "visible minorities"). The risk of discrimination has been one of the reasons the United Kingston has considered systematic screening of SCD rather than maintaining the selective screening.

The risk of non-paternity: Testing parents of carrier infants identified by the neonatal screening raises the risk of discovering that the putative father is not the biological father ("non-paternity").[17]

5.2   *Cultural Aspects*

The general public is better informed about progress in genetics (newspapers, TV programmes, fundraising programmes (such as the Telethon in France, websites, etc), but very little information is provided about neonatal screening.

It is often difficult for uneducated parents to distinguish between carrier and disease status.

In population subgroups where there is a high incidence of ("traditional") consanguinity, explaining genetic risk is often difficult. In the case of SCD screening, most African men consider that blood-related illnesses are transmitted exclusively by women. This belief is so widespread that many women who are aware of their carrier status prefer to keep silent in order to be able to get married.[18]

How the disclosure of genetic status would change their future reproductive decisions requires further studies, because cultural and religious reasons may have a strong negative influence on requests for prenatal diagnosis.

6.   THE CHOICES

Options include employing tests that do not identify carrier status, if available; identifying acceptable ways of disclosing carrier status; or identifying acceptable ways of not disclosing carrier status.

Nowadays, withholding the information is not conceivable, but it may be difficult to justify giving information on infants' carrier status to parents without also providing counselling. However, genetic counselling would have to be provided to several thousand couples each year; this would require specific infrastructure and financing which has not been granted.

For SCD, since there are no screening tests available for sickle cell disorders that do not identify carrier status, the communication of carrier results is the responsibility of the "regional reference physician." A booklet for parents has also been designed to explain the meaning of "heterozygote."

In a family with a confirmed CF newborn, genetic counselling and a mutation search are reasonable, but more questionable when a carrier is detected. The main question concerns the "pathological" significance of mutations which are included in the kit in use. In fact, panels are based on the relative frequency of *CFTR* mutations in the population, and not on genotype-phenotype relationships. In addition, the performance of the strategy to detect carriers is poor, since only 1% to 2% of carriers are identified. For these reasons, alternative strategies which do not identify carrier status are explored. A 2-tiered CF newborn screening strategy with a first test for elevated immunoreactive trypsinogen (IRT) with subsequent analysis for elevated pancreatitis-associated protein (PAP) has been evaluated on 205 000 newborns.[19] The diagnosis is confirmed by the sweat test. Such a strategy screens for a phenotype and no longer for a genotype.

7. CONCLUSION

For SCD, although targeting raises some difficulties, there is no trend in France to consider universal neonatal screening. Carrier status is communicated to the referring physician.

For CF, detection of heterozygotes involves the same process. But efforts must be made to get a better definition of the disease and to be able to give better information on the significance of the risks. If not, new strategies which do not identify carrier status would be preferable.

This is relevant to future developments in molecular genetics, which may place health services under increasing pressure to test for a wide range of genetic conditions in early life, many of which have no immediate implications for health.

TABLE 1: DESCRIPTION OF THE FRENCH NEONATAL SCREENING PROGRAMME

DISEASE	START OF SCREENING*	METHOD	NUMBER OF NEWBORNS TESTED (NUMBER OF CONFIRMED CASES) **
PKU	1967	Phenylalanine measurement	23 664 120 (1439)
CH	1978	TSH measurement	20 589 854 (5786)
CAH	1981	17OH progesterone measurement	8 785 694 (574)
SCD	1985 oversea departments	haemoglobin profile by isoelectric focusing	478 220 (1073) (systematic screening)
	2000 metropolitan France		1 364 573 (1674) (targeted screening of population at risk for SCD)
CF	2002	IRT measurement and DNA testing (panel of 30 mutations) for samples with elevated IRT (>percentile 99.5)	1 928 911 (428)

*   Year of the start of the progressive implementation of screening in the 21 regions

**   Data from AFDPHE at the end of 2004

TABLE 2: CHARACTERISTICS OF SICKLE CELL DISEASE AND CYSTIC FIBROSIS IN REGARD TO
THEIR NEONATAL SCREENING AND THE POSSIBILITY OF IDENTIFYING CARRIERS

	Sickle Cell Disease	Cystic Fibrosis
Genetics	Autosomal recessive	Autosomal recessive
Incidence:		
West Indies & French Guiana	1/278	?*
African immigrants	1/50 to 1/200	1/15000
Caucasian	rare (1/150,000)	1/2500 to 1/4500
Symptoms	Severe anemia, susceptibility to infections, acute painful episodes (hemolytic, splenic sequestration, aplastic crisis)	Chronic pulmonary disease, exocrine pancreatic insufficiency (a wide spectrum of clinical variability exists)
Treatment	Daily antibiotic prophylaxis, immunization, parental education	(No truly curative treatment) digestive enzyme supplements, physiotherapy, antibiotics
Main advantage of early recognition	Prevent bacterial infections which can become life threatening, reduction of pneumococcal infections and of the mortality related to splenic sequestration	Eliminates the "diagnostic odyssey" (period of uncertainty) that generally precedes clinical diagnosis
Definition of the disease	"major sickle cell syndrome" SS, SC, S/β thalassemia	"There is no absolute definition of CF"[20] What is the significance of "mild mutations" (i.e. R117H)?
Screening method	Isoelectric focusing, HPLC, … "genetic information"	IRT + DNA testing "genetic test" high allelic heterogeneity (1415 mutations listed on Nov. 2005)
Cost per test	2.64 €	1.57 € /IRT 122 €/DNA test
Carrier recognition	Almost 100% (A/S, A/C) 4% of tested newborns	1/2000 (*vs* 1/25)
Alternative option without looking at the genotype	No method!	Inclusion of PAP assay
Impact of "BioEthics Law"	None (Because haemoglobin electrophoresis is in "routine" use in labs, this parameter is not considered a "genetic test")	Yes
Impact of the need for written consent on the coverage		0.2% of refusal**
Communication of carrier status to other members of the family	The physician has to inform the patient about the risk of not disclosing the result to the other members of the family. But it is the patient's right to disclose or not disclose his or her genetic status	

*   Systematic screening not yet implemented in overseas departments
**   Data from the screening centre of Lille (2002-2005, 277000 Guthrie cards)

References

1. J. L. Dhondt et al., *Neonatal Screening in France*, 2 Screening 77 (1993).
2. J. L. Dhondt et al., *Results of Pilot Screening Activities in the French Neonatal Screening–Cystic Fibrosis, Congenital Adrenal Hyperplasia and Sickle Cell Disease*, 2 Screening 87 (1993).
3. J. P. Farriaux et al., *Neonatal Screening for Cystic Fibrosis: France Rises to the Challenge*, 26 J. Inher. Met. Dis. 729 (2003).
4. A. M. Comeau et al., *Population-Based Newborn Screening for Genetic Disorders when Multiple Mutation DNA Testing is Incorporated: A Cystic Fibrosis Newborn Screening Model Demonstrating Increased Sensitivity but More Carrier Detections*, 113 Pediatrics 1573 (2004).
5. K. G. Monaghan et al., *Preconception and Prenatal Cystic Fibrosis Carrier Screening of African Americans Reveals Unanticipated Frequencies for Specific Mutations*, 6 Genet Med. 41 (2004).
6. Law No. 78-17 of January 6, 1978, Journal Officiel de la République Française [J.O.] [Official Gazette of France], July 18, 1978.
7. Law No. 88-1138 of December 20, 1988, Journal Officiel de la République Française [J.O.] [Official Gazette of France], December 22, 1988.
8. Law No. 94-653 of July 29, 1994, Journal Officiel de la République Française [J.O.] [Official Gazette of France], July 30, 1994.
9. Law No. 94-654 of July 29, 1994, Journal Officiel de la République Française [J.O.] [Official Gazette of France], July 30, 1994.
10. Law No. 96- 452 of May 28, 1996, Journal Officiel de la République Française [J.O.] [Official Gazette of France], May 29, 1996.
11. Decree No. 2000-570 of June 23, 2000, Journal Officiel de la République Française [J.O.] [Official Gazette of France], June 27, 2000.
12. Law No. 2004-800 of August 6, 2004, Journal Officiel de la République Française [J.O.] [Official Gazette of France], August 7, 2004.
13. J. L. Dhondt & J. P. Farriaux, *Impact of French Legislation on Neonatal Screening, in* Human DNA: Law and Policy 285 (B. M. Knoppers ed., 1997).
14. Law No. 2004-800.
15. E. P. Parsons et al., *Implications of Carrier Identification in Newborn Screening for Cystic Fibrosis*, 88 Arch. Dis. Child. Fetal Neonatal Ed. F467 (2003).
16. S. Oliver et al., *Disclosing to Parents Newborn Carrier Status Identified by Routine Blood Spot Screening*, 4 Cochrane Database Syst. Rev. CD003859 (2004).
17. L. Laird et al., *Neonatal Screening for Sickle Cell Disorders: What About the Carrier Infants?* 313 B. M. J. 407 (1996).
18. M. De Montalembert et al., *Ethical Aspects of Neonatal Screening for Sickle Cell Disease in Western European Countries,* 94 Acta Paediatr. 528 (2005).
19. J. Sarles et al., *Combining Immunoreactive Trypsinogen and Pancreatitis-Associated Protein Assays. A Method of Newborn Screening for Cystic Fibrosis that Avoids DNA Analysis,*147 J. Pediatr. 302 (2005).
20. B. J. Rosenstein, *What is a Cystic Fibrosis Diagnosis?* 19 Clin. Chest. Med. 423 (1998).

Carrier Detection in Newborns: Should it be Discovered? Should it be Disclosed? Lessons from Sickle Cell Anemia and Cystic Fibrosis Screening in the United States

Lainie FRIEDMAN ROSS

University of Chicago, Department of Pediatrics, USA

CAROLYN and MATTHEW BUCKSBAUM PROFESSOR

University of Chicago, Department of Pediatrics, USA

1.   INTRODUCTION

In 1973, Norm Fost and Michael Kaback wrote an article for the journal *Pediatrics* entitled, "Why do Sickle Screening in Children? The Trait is the Issue."[1] Their article was a critique of sickle cell screening programmes. They cited a Massachusetts law that required all children to have sickle cell screening to identify "sickle trait" prior to entry into public school.[2] Fost and Kaback expressed concern about diagnosing thousands of children as heterozygote carriers when there was still a lot of confusion both in the medical and lay community about its significance. Fost and Kaback argued that there are advantages in diagnosing individuals with sickle cell disease, but this could be done by screening children over one year of age for anemia first. By using haemoglobin as the primary screening method rather than haemoglobin electrophoresis, one could avoid detecting most carriers.[3] They argued that screening programmes that detect carriers should be initiated only *after* pilot studies have resolved the issue of the clinical significance of carrier status.

The concerns expressed by Fost and Kaback were put aside in 1986 with the discovery that penicillin prophylaxis could prevent serious morbidity and even mortality in infants and young children with sickle cell anemia (hereafter SCA).[4] Now a newborn diagnosis was critically important, and the problem of discovering sickle cell carrier status was accepted as a foreseen but unintended consequence.

The issue of carrier discovery in newborns re-emerged in the United States in the early 1990s, following the 1989 cloning of the cystic fibrosis transmembrane conductance regulator (CFTR) gene by John Riordan, Francis Collins and Lap-Chee Tsui.[5] At the time, newborn screening (NBS) for cystic fibrosis was being piloted in a few states by measuring immunoreactive trypsinogen (IRT),[6] but this method required two samples. A two-tiered screening IRT/DNA protocol was developed and implemented quickly.[7] Samples were tested for elevated levels of IRT. Those above a certain threshold were then tested for the ΔF508 mutation, the most common mutation found in the CFTR gene. This protocol allows for screening with only one sample, but it diagnoses some heterozygote carriers. And in fact, as more mutations are discovered and added to the newborn screening panel, an even larger number of carriers are able to be diagnosed.[8]

This chapter explores issues of detection and disclosure of carrier status in sickle cell anemia (SCA) and cystic fibrosis (CF) as a routine component of NBS in the United States (US). It argues that detection and disclosure must be understood within the social context in which NBS occurs. That is, the practices and policies of detecting and disclosing carriers as part of NBS should not be examined in isolation, but rather, in conjunction with other genetic testing programmes like antenatal and population carrier programmes for SCA and CF.

2.   SICKLE CELL ANEMIA SCREENING PROGRAMMES

Sickle cell anemia (SCA) is an autosomal recessive disease characterized by anemia and vaso-occusive events. Although it is due to a single point mutation, there is wide variability in symptomatology. In the US, 10% of Black Americans are carriers and 0.3% have sickle cell anemia.[9] It is also seen in other communities (e.g., Mediterranean families). Carriers are usually asymptomatic although they may have slight increased risk of sudden death in anaerobic situations (e.g., high altitudes and heat exhaustion).[10] The frequency of the allele is explained in part because carriers are somewhat protected against malaria.

The sickledex screen was developed by Greenberg in 1972.[11] It was a solubility test but could not distinguish carriers from homozygotes. Given the high morbidity and mortality from sickle cell anemia in the US Black population, and a strong desire to "do something,"[12] the National Sickle Cell Anemia and Control Act was passed that same year.[13] Throughout the 1970s, many pilot programmes were funded by the National Sickle Cell Anemia and Control Act and programmes set up around the country focused on population screening and screening pregnant women and their partners. Many of these programmes were failures because of the misunderstanding and confusion by physicians and the wider community between carriers and those who are affected.[14] They were begun prematurely, with little regard for potential harmful effects; public education was not provided; and counseling was insufficient.[15]

Within a year, Garrick et al. would describe an eletrophoresis methodology which could distinguish trait from disease and which could be done on filter paper collected for newborn screening for PKU.[16] However, newborn screening for sickle cell disease would not gain in popularity for another decade. In 1985, only 7 states were screening for haemoglobinopathies along with PKU.[17] But in 1986, Gaston et al. showed that penicillin prophylaxis decreased morbidity and mortality of children with sickle cell anemia[18] and in 1987, a consensus panel was held by the Centers for Disease Control and Prevention (CDC) recommending newborn screening for sickle cell anemia.[19] By 1990, 29 states offered some type of NBS for SCA, but it was often targeted and not universal.[20] In 2006, 49 states plus the District of Columbia provide universal NBS. Testing is required in New Hampshire, but has not been implemented.[21]

NBS programmes use the haemoglobin electrophoresis method which detects all affected individuals with haemoglobin SS disease as well as other haemoglobinopathies, some of variable significance. Haemoglobin electrophoresis also detects carriers of one haemoglobin S allele (heterozygote carriers). In the 1970s and 1980s, there was wide variability in disclosing carrier results. For example, New York routinely disclosed carrier findings,[22] but California did not.[23] Genetic counseling was offered but frequently not provided.[24] The President's commission in 1983 stated that newborn screening should not be done primarily to determine parental carrier results.[25] However, if carriers are identified, parents have a right to know and to be counseled about its significance. This was reaffirmed in the Institute of Medicine Report, *Assessing Genetic Risks* in 1994.[26] Today, all US states disclose carrier results with variable degrees of genetic counseling provided.

3.   CYSTIC FIBROSIS

Cystic fibrosis is an autosomal recessive disease characterized by gastrointestinal disease (pancreatic insufficiency and malnutrition) and pulmonary disease. Approximately 4% of Caucasians are carriers, and about .04% have CF.[27] Over 1000 mutations have been identified although at least one copy of the ΔF508 mutation is found in over 70% of Caucasians with CF.[28] CF exists in other ethnic communities, but is less common and is often associated with different mutations.[29] While there is good genotypic-phenotypic correlation for gastrointestinal symptoms, there is poor genotypic-phenotypic correlation for pulmonary symptoms.[30] In general, carriers are asymptomatic although they may be at greater risk for sinusitis and asthma.[31] There is some research in mice that suggests that CF heterozygotes may have some protection from cholera.[32]

While the initial SCA screening programmes were carrier screening programmes, CF screening began as a newborn programme with the development of an immunoreactive trypsinogen (IRT) assay that could be performed on dried blood spots.[33] Although Colorado began a service programme almost immediately,[34] a CDC workshop in 1983 concluded that such screening would be premature as there were not enough data.[35] In 1985, Wisconsin began randomized controlled trials using IRT/IRT method.[36] After Collins et al. cloned the CFTR gene in 1989, the Wisconsin research team revised their protocol to employ the two-tiered IRT/DNA method.

Population and antenatal programmes also became possible with the discovery of the ΔF508 mutation. However, many studies showed low uptake, even when offered for free.[37] But in 2001, the American College of Obstetrics and Gynecology (ACOG) recommended universal carrier screening of women antenatally or preconception for 25 mutations.[38] The result was a large increase in the number of women screened prenatally.

Uptake for newborn screening for CF got a large push when the American College of Medical Genetics (ACMG) recommended the inclusion of CF in their uniform panel (2004).[39] This came on the heels of a CDC conference in 2003 where Botkin concluded that CF newborn screening was justified although it should be dependent on state resources.[40] In 1990, only 3 states were screening for CF[41] and this was unchanged 7 years later.[42] Today, in contrast, 15 states screen for CF as part of universal screening, and it is mandated in others but not yet implemented.[43]

There are a number of different methodologies to screen newborns for CF. The initial method employed IRT screening with repeat IRT testing two weeks later in those who were in the highest range, followed by a confirmatory sweat test in those who had two positive IRT screens. With the discovery of the most common CF mutation, there was some support for using a single sample two-tiered IRT/DNA methodology. As more and more mutations were discovered, IRT/DNA became more accurate, but more carriers were detected as well.[44] Some concern has been expressed about consent and confidentiality with respect to gene-based testing.[45]

Another way to perform CF screening with a single sample is to do IRT/PAP (pancreatic associated protein). This has the advantage of a single sample without the need for genetic testing. It is being tested in France, but not in the US.[46] This may be due to the fact that countries like France require written informed consent for any gene-based testing, even if performed within a newborn screening programme.[47]

For those states that use the IRT/DNA method, some carriers of CF will be picked up. How many carriers are discovered depends on the number of mutations included in the newborn screening panel as well as the IRT cut-off. No current newborn screening programme for CF, however, identifies all infants who are carriers, because only children who have an elevated IRT undergo genetic testing. Thus, the IRT/DNA method expects to diagnose only one in 200 carriers (rather than one-in-30 in the general population) that would be discovered if DNA-based screening were first-tier. Carriers identified by the IRT/DNA method undergo sweat testing to ensure that they do not have CF due to a second, less common mutation.

It is also the case that some children who have two CF mutations will not be identified by newborn screening because DNA analysis is only performed on those with an elevated IRT. While this would be problematic for a condition like PKU in which maximizing sensitivity is the essential goal to ensure that all affected children began dietary treatment early, this approach has not been used in the current US DNA-based screening programmes.[48] Rather, it is not evident that it is critical to identify those with mild CF mutations who may not be symptomatic for decades,[49] particularly those who have a pancreatic sufficient form of CF because early therapy has been shown to be most critical in treating failure to thrive.[50] In fact, some argue against the IRT/DNA method precisely because it picks up some gentoypes with known mild disease.[51]

States and countries that use the IRT/DNA methodology differ on which alleles and how many alleles are included in the newborn screening protocol. In general, the IRT/DNA methodology is more accurate in Caucasian populations but less accurate in other ethnic populations.[52] To achieve greater equity in diagnosing CF in all populations would require either that the DNA screen include more mutations or that a different screening methodology be employed.

4.   COMPARATIVE ANALYSIS: THE NEED FOR DISCOVERY

The historical developments that led to screening for SCA and CF involved divergent paths. Whereas SCA began as a programme to identify carriers in the general population, CF screening began as a programme to identify affected newborns using a non-genetic test (see Table 1). With the development of haemoglobin electrophoresis and the ability to test samples collected on filter paper, NBS for SCA became possible, but it did not grow in popularity until an effective preventive therapy (penicillin prophylaxis) made it useful. When the gene for CF was discovered, ΔF508 screening was incorporated into NBS protocols, and population and prenatal carrier programmes were developed. Population programmes traditionally had low uptake in the US.[53] Prenatal uptake programmes have higher uptake but were not uniformly offered.[54] Uptake increased substantially, however, when the American College of Obstetrics and Gynecology declared CF carrier testing to be the standard of care in 2001.[55]

TABLE 1: ORDER IN WHICH SCREENING PROGRAMMES WERE DEVELOPED FOR SICKLE CELL ANEMIA AND CYSTIC FIBROSIS

	POPULATION SCREENING	ANTENATAL SCREENING	NEWBORN SCREENING
SCA	1st	2nd	3rd
CF	2nd	3rd	1st

When discussing whether or not to discover carrier infants in NBS programmes, one must consider that other modes of identifying carriers exist. Most pregnant women are offered carrier testing for SCA based on ethnicity and CF almost universally.[56] Thus, the value of identifying heterozygote newborns in order to initiate cascade screening of parents and relatives is not necessary because these adults are routinely offered antenatal carrier testing. Since carrier identification does not have any meaningful impact on the health and well-being of the infants themselves, disclosure runs the risk of confusing parents without providing much benefit to the family.[57]

That said, newborn screening for SCA identifies 100% of carriers. In contrast, NBS for CF can be done using methodologies that do not discover carriers. Even when using the IRT/DNA method, many carriers are not discovered. The only way to discover all carriers would be to employ a first-tier DNA screen, but this does not offer clinical benefit at this time.

5.   COMPARATIVE ANALYSIS: THE NEED FOR DISCLOSURE?

Given that current testing methodologies require discovery of heterozygote sickle cell carriers; and that some CF programmes use a protocol that discovers heterozygote CF carriers, the question remains whether or not there is a need to disclose this information.

There is consensus that carrier information about a newborn is information about the child that belongs to the child, but the child is too young to be given this information. This leaves two options. One is to leave the information with the state or in the child's medical record to be accessed when the child is an adult. The problem with this solution is it assumes that third parties should decide how to store this information and when to disclose it, or that children as adults will know how to access it. It also denies that parents are presumed to be the surrogate for the child.

The second option is to argue that the parents are the appropriate surrogate for the child's health care until the child is able to make his or her own health care decisions and as such, the parents have a right and obligation to be informed of the child's carrier status. To the extent that carrier information has health risks, one could argue that parents have a right to this information about their child. However, when carrier information entails no health risks to the child and is really only information about a child's (and a parent's) reproductive risks, the parental claim of a right to know is less compelling.

What are the benefits of parental disclosure? The first is that it gives the parents the opportunity to pass this information onto the child at a developmentally appropriate time and in a developmentally appropriate manner. Of course, this may backfire if the parents misunderstand or forget the information and pass on misinformation, or if they confuse the information over time, or if they choose not to reveal it.

A second potential benefit is that discovery and disclosure of infant carriers may decrease disparities because not all women have access to antenatal testing. In addition, antenatal screening generally focuses on screening women, whereas a carrier infant may point to a carrier father. The strongest objection to these considerations is that using the infant to gain reproductive information for the parents is not justified when the parents themselves can be tested. In addition, it is the case that not all women and couples want to be aware of their carrier status. Thus, identification and disclosure of an infant's carrier status denies the parents their right not to know. And in fact, it also denies the child his or her right not to know.

The right not to know is an important theme in genetics literature.[58] The argument is strongest when the genetic information yields future health risks about a person for a highly penetrant, untreatable condition like Huntington disease.[59] Once known, the information cannot be unknown and the individual must decide how to use the information, not whether to use it. In contrast, when the genetic information is about carrier status for an autosomal recessive condition, the information is mainly about a risk that is only relevant in reproductive decision-making. Although the information cannot be unknown, the individual or couple can decide whether to use the information and how to use it. This is not to deny that carrier information can be anxiety-provoking and undesired, only to say that its discovery is more easily justified than the unrequested presymptomatic information about a person's future health. Discovery of carrier information can be justified if it is an unavoidable by-product of diagnosing affected children for whom early diagnosis reduces morbidity and mortality.

In addition to violating the parents' right not to know, there are other potential risks of giving carrier information to the parents. One concern is that it may adversely affect the parent-child relationship,[60] although there are no long-term data to confirm this. A second concern is that this label may lead to discrimination or stigma against the child.[61] Here, we do know that carriers of sickle cell anemia have experienced insurance discrimination and stigma.[62] But with the ever expansion of genetic carrier information, we will all be carriers of numerous autosomal recessive conditions, some potentially lethal. Whether this will dissipate discrimination is as yet unknown.

A third concern is that individuals will misunderstand the meaning of false positives and carrier status.[63] In 1974, researchers described heterozygote carriers of SCA who were treated differently by their families after diagnosis. They referred to this problem as "sickle cell nondisease."[64] Likewise, in the Wisconsin study on newborn screening for CF, about 5-10% of parents counseled that their child was a carrier of one CF gene, were convinced that their child had "a touch of CF."[65]

After weighing the pros and cons of carrier disclosure, I support disclosure. Once the information is obtained, it belongs to someone, and I believe it belongs to the child and hence the parents must be the repository of this information. This is particularly true given that this information is obtained in the US in mandatory universal programmes without informed consent.

Although I have argued that the balance favours disclosure, let me add three caveats. First, carrier detection should not be the sole reason for NBS. Second, NBS methodologies that can avoid carrier determinations should be considered, particularly if screening is done in a mandatory programme. In that vein, I support the use of IRT/IRT or IRT/PAP screening rather than IRT/DNA for CF at this time. Third, if consent were necessary for newborn screening, one could envision that health care providers would need to explain to parents that carrier information is an incidental finding of state-sponsored NBS and that the preferred policy is not to disclose these findings as they are not clinically relevant to the child. Under an informed consent framework, parents would be encouraged to consent to non-disclosure, although they could insist upon disclosure. Thus, the process of informed consent allows for selective disclosure of information which is more respectful of parental autonomy and the child's right to privacy.

6. CONCLUDING THOUGHTS

In conclusion, let me emphasize three points. First, NBS should be undertaken with the primary goal of promoting the child's medical well-being. NBS carrier detection and disclosure for SCA and CF do not serve this function. Second, NBS must be understood within the political and social contexts in which it is offered. Newborn screening is mandatory in 48 of 50 states, and consent is not obtained. To that end, in the US, NBS screening programmes that can avoid the detection of carrier information are and ought to be preferred. Furthermore, in the US, antenatal screening for CF is recommended universally[66] and antenatal screening for SCA is recommended based on ethnicity.[67] For many, then, NBS carrier detection and disclosure will be either redundant information or information that they choose not to procure.

Third, although NBS in the US is currently designed as a mandatory public health programme, there is movement to expand NBS beyond the traditional public health criteria enumerated by Wilson and Jungner.[68] Supporters of expanded NBS seek to include conditions for which there are no known effective therapies and to include conditions for which the natural history is not well understood.[69] This movement provides a powerful reason to re-evaluate current NBS policy and practice that exclude parental decision making. Data show that most parents will consent to expanded screening, whether or not the conditions meet the Wilson and Jungner criteria.[70] The informed consent conversation should include a discussion of the benefits and risks of screening, including the potential discovery of unintended byproducts like heterozygote carriers. Ideally, the informed consent conversation would include an explanation of why the discovery of carrier information is not relevant to the child *qua* child and why it should not be revealed even when it is discovered as an unintended byproduct of NBS. But until consent is routine, the unintended detection of carrier status in NBS must be disclosed to the parents in their role as the child's surrogate.

REFERENCES

1. N. Fost & M. M. Kaback, *Why Do Sickle Cell Screening in Children? The Trait Is the Issue*, 51 PEDIATRICS 742 (1973).
2. *Id.*
3. *Id.*
4. M. H. Gaston et al., *Prophylaxis with Oral Penicillin in Children with Sickle Cell Anemia. A Randomized Trial*, 314 N. ENG. J. MED. 1493 (1986).
5. J. R. Riordan et al., *Identification of the Cystic Fibrosis Gene: Cloning and Characterization of Complementary DNA*, 245 SCIENCE 1066 (1989); B. Kerem et al., *Identification of the Cystic Fibrosis Gene: Genetic Analysis*, 245 SCIENCE 1073 (1989).
6. K. B. Hammond et al., *Efficacy of Statewide Neonatal Screening for Cystic Fibrosis by Assay of Trypsinogen Concentrations*, 325 N. ENGL. J. MED. 769 (1991); R. G. Gregg et al., *Newborn Screening for Cystic Fibrosis in Wisconsin: Comparison of Biochemical and Molecular Methods*, 99 PEDIATRICS 819 (1997).
7. Gregg et al., *supra* note 6.
8. A. M. Comeau et al., *Population-Based Newborn Screening for Genetic Disorders when Multiple Mutation DNA Testing Is Incorporated: A Cystic Fibrosis Newborn Screening Model Demonstrating Increased Sensitivity but more Carrier Detections*, 113 PEDIATRICS 1573 (2004).
9. M. H. Steinberg, *Management of Sickle Cell Disease*, 340 N. ENG. J. MED. 1021 (1999).
10. J. A. Kark et al., *Sickle-Cell Trait as a Risk Factor for Sudden Death in Physical Training*, 317 N. ENG. J. MED. 781 (1987); D. P. Wirthwein et al., *Death Due to Microvascular Occlusion in Sickle-Cell Trait Following Physical Exertion*, 46 J. FORENSIC SCI. 399 (2001).
11. M. S. Greenberg et al., *A Simple and Inexpensive Screening Test for Sickle Hemoglobin*, 286 N. ENG. J. MED. 1143 (1972).
12. Fost & Kaback, *supra* note 1.
13. *National Sickle Cell Anemia Control Act*, Public Law no. 92-294, 86 Statute 138, 1972 (May 16, 1972).
14. Institute of Society, Ethics, and the Life Sciences, *Ethical and Social Issues in Screening for Genetic Disease*, 286 N. ENG. J. MED. 1129 (1972); P. Reilly, Genetics, Law and Social Policy (1977); Committee for the Study of Inborn

Errors of Metabolism, National Research Council, Genetic Screening: Programs, Principles, and Research (1975).

15. Institute of Society, Ethics, and the Life Sciences, *supra* note 14; Reilly, *supra* note 14; Committee for the Study of Inborn Errors of Metabolism, National Research Council, *supra* note 14.

16. M. D. Garrick et al., *Sickle-Cell Anemia and Other Hemoglobinopathies. Procedures and Strategy for Screening Employing Spots of Blood on Filter Paper as Specimens*, 288 N. ENG. J. MED. 1265 (1973).

17. Reilly, *supra* note 14.

18. Gaston et al., *supra* note 4.

19. Consensus Conference, *Newborn Screening for Sickle Cell Disease and other Hemoglobinopathies*, 258 J. A. M. A. 1205 (1987).

20. P. T. Rowley, *Newborn Screening for Hemoglobinopathies*, 14 SEMINARS IN PERINATAOLOGY 483 (1990).

21. National Newborn Screening and Genetics Resource, National Newborn Screening Status Report, http://genes-r-us.uthscsa.edu/nbsdisorders.pdf.

22. Reilly, *supra* note 14.

23. D. Powars, *Diagnosis at Birth Improves Survival of Children with Sickle Cell Anemia*, 83 PEDIATRICS 830 (1989).

24. Reilly, *supra* note 14.

25. President's Commission for the Study of Ethical Problems in Medicine and Biomedical and Behavioral Research, Screening and Counseling for Genetic Conditions: The Ethical, Social, and Legal Implications of Genetic Screening, Counseling, and Education Programs (1983).

26. L. B. Andrews et al., eds. (on behalf of the Committee on Assessing Genetic Risks, Institute of Medicine), Assessing Genetic Risks: Implications for Health and Social Policy (1994).

27. American College of Genetics, Committee on Genetics, *Update on Carrier Screening for Cystic Fibrosis*, 106 OBSTET. GYNECOL. 1465 (2005).

28. S. D. Grosse et al., *CDC. Newborn Screening for Cystic Fibrosis: Evaluation of Benefits and Risks and Recommendations for State Newborn Screening Programs*, 55 MORBIDITY & MORTALITY WEEKLY REPORT. RECOMMENDATIONS & REPORTS (RR-13) 1 (2004).

29. American College of Genetics, Committee on Genetics, *supra* note 27; Grosse et al., *supra* note 28.

30. F. Salvatore et al., *Genotype-Phenotype Correlation in Cystic Fibrosis: The Role of Modifier Genes*, 111 AM. J. MED. GENET. 88 (2002).

31. M. Dahl et al., *DeltaF508 Heterozygosity in Cystic Fibrosis and Susceptibility to Asthma*, 351 LANCET 1911 (1998); V. Raman et al., *Increased Prevalence of Mutations in the Cystic Fibrosis Transmembrane Conductance Regulator in Children with Chronic Rhinosinusitis*, 109 PEDIATRICS E13 (2002).

32. J. Bertranpetit & F. Calafell, *Genetic and Geographical Variability in Cystic Fibrosis: Evolutionary Considerations*, 197 CIBA FOUNDATION SYMPOSIUM 97 (1996).

33. J. R. Crossley et al., *Dried-Blood Spot Screening for Cystic Fibrosis in the Newborn*, 1 LANCET 472 (1979).

34. Hammond et al., *supra* note 6.

35. Cystic Fibrosis Foundation, *Neonatal Screening for Cystic Fibrosis: Position Paper*, 72 PEDIATRICS 741 (1983).

36. Gregg et al., *supra* note 6.

37. E. W. Clayton et al., *Lack of Interest by Nonpregnant Couples in Population-Based Cystic Fibrosis Carrier Screening*, 58 AM. J. HUM. GENET. 617 (1996); L. Henneman et al., *Participation in Preconceptional Carrier Couple Screening: Characteristics, Attitudes, and Knowledge of Both Partners*, 38 J. MED. GENET. 695 (2001); S. Loader et al., *Cystic Fibrosis Carrier Population Screening in the Primary Care Setting*, 59 AM. J. HUM. GENET. 234 (1996).

38. Loader et al., *supra* note 37.

39. American College of Medical Genetics/Health Resources and Services Administration, Newborn Screening: Toward a Uniform Screening Panel and System (2004), ftp://ftp.hrsa.gov/mchb/genetics/screeningdraftforcomment.pdf.

40. Grosse et al., *supra* note 28.

41. Andrews et al., *supra* note 26.

42. Grosse et al., *supra* note 28.

43. National Newborn Screening and Genetics Resource, *supra* note 21.

44. Comeau et al., *supra* note 8.

45. J. Sarles et al., *Combining Immunoreactive Trypsinogen and Pancreatitis-Associated Protein Assays, a Method of Newborn Screening for Cystic Fibrosis that Avoids DNA Analysis*, 147 J. PEDIATR. 302 (2006); N. S. Green & K. A. Pass, *Neonatal Screening by DNA Microarray: Spots and Chips*, 6 NAT. REV. GENET. 147 (2005).

46. J. Sarles et al., *Blood Concentrations of Pancreatitis Associated Protein in Neonates: Relevance to Neonatal Screening for Cystic Fibrosis*, 80 ARCH. DIS. CHILD. FETAL NEONATAL ED. F118 (1999).

47. J. Sarles et al., *supra* note 45; S. Barthellemy et al., *Évaluation sur 47 213 enfants d'une stratégie de dépistage néonatal de la mucoviscidose associant les dosages de pancreatitis*, 8 ARCH. PEDIATR. 275 (2001); J. L. Dhondt, *Implementation of Informed Consent for a Cystic Fibrosis Newborn Screening Program in France: Low Refusal Rates for Optional Testing*, 147 J. PEDIATRICS (supplement) S106 (2005).

48. Comeau et al., *supra* note 8; B. S. Wilfond & S. E. Gollust, *Policy Issues for Expanding Newborn Screening Programs: The Cystic Fibrosis Newborn Screening Experience in the United States*, 146 J. PEDIATRICS 668 (2005).

49. Sarles et al., *supra* note 45; J. S. Wagener et al., *Update on Newborn Screening for Cystic Fibrosis*, 10 CUR. OPIN. PULM. MED. 500 (2004).

50. Grosse et al., *supra* note 28; Wagener et al., *supra* note 49.

51. Sarles et al., *supra* note 45.

52. Green & Pass, *supra* note 45; K. G. Monaghan et al., *Preconception and Prenatal Cystic Fibrosis Carrier Screening of African Americans Reveals Unanticipated Frequencies for Specific Mutations*, 6 GENET. MED. 141 (2004).

53. Clayton et al., *supra* note 37; P. T. Rowley et al., *Cystic Fibrosis Carrier Population Screening: A Review*, 1 GENET. TEST. 53 (1997).

54. Loader et al., *supra* note 37; Rowley et al., *supra* note 53.

55. American College of Obstetricians and Gynecologists & American College of Medical Genetics, Preconception and Prenatal Carrier Screening for Cystic Fibrosis: Clinical and Laboratory Guidelines (2001).

56. American College of Obstetrics and Gynecology Committee, *ACOG Practice Bulletin. Clinical Management Guidelines for Obstetrician-Gynecologists: Hemoglobinopathies in Pregnancy*, 106 OBSTET. GYNECOL. 203 (2005); Andrews et al., *supra* note 26.

57. M. L. Hampton, *Sickle Cell "Nondisease": A Potentially Serious Public Health Problem*, 128 AM. J. DIS. CHILD. 58 (1974); D. J. Ciske et al., *Genetic Counseling and Neonatal Screening for Cystic Fibrosis: An Assessment of the Communication Process*, 107 PEDIATRICS 699 (2001); E. H. Mischler et al., *Cystic Fibrosis Newborn Screening: Impact on Reproductive Behavior and Implications for Genetic Counseling*, 102 PEDIATRICS 44 (1998).

58. R. Chadwick et al., The Right to Know and the Right Not to Know (1997); T. Takala, *The right to Genetic Ignorance Confirmed*, 13 BIOETHICS 288 (1999); J. Harris & K. Keywood, *Ignorance, Information and Autonomy*, 22 THEOR. MED. 415 (2001); K. A. Quaid et al., *Knowledge, Attitude, and the Decision to be Tested for Huntington's Disease*, 36 CLIN. GENET. 431 (1989).

59. Quaid et al., *supra* note 58.

60. Andrews et al., *supra* note 26; Working Party of the Clinical Genetics Society (UK), *The Genetic Testing of Children*, 31 J. MED. GENET. 785 (1994); American Society of Human Genetics/American College of Medical Genetics, *Points to Consider: Ethical, Legal, and Psychosocial Implications of Genetic Testing in Children and Adolescents*, 57 AM. J. HUM. GENET. 1233 (1995); R. M. Nelson et al., *Committee on Bioethics. Ethical Issues with Genetic Testing in Pediatrics*, 197 PEDIATRICS 1451 (2001).

61. President's Commission for the Study of Ethical Problems in Medicine and Biomedical and Behavioral Research, *supra* note 25; Andrews et al., *supra* note 26; Party of the Clinical Genetics Society (UK), *supra* note 60; American Society of Human Genetics/American College of Medical Genetics, *supra* note 60; Nelson et al., *supra* note 60; R. H. Kenen & R. M. Schmidt, *Stigmatization of Carrier Status: Social Implications of Heterozygote Genetic Screening Programs*, 68 AM. J. PUBLIC HEALTH 1116 (1978).

62. Andrews et al., *supra* note 26.

63. Hampton, *supra* note 57; M. B. Rothenberg & E. M. Sills, *Iatrogenesis: The PKU Anxiety Syndrome*, 7 J. AM. ACAD. CHILD PSYCHIATRY 689 (1968); A. Tluczek et al., *Parents' Knowledge of Neonatal Screening and Response to False-Positive Cystic Fibrosis Testing*, 13 J. DEV. BEHAV. PEDIATR. 181 (1992).

64. Hampton, *supra* note 57.

65. Tluczek et al., *supra* note 63.

66. American College of Genetics, Committee on Genetics, *supra* note 27.

67. American College of Obstetrics and Gynecology Committee, *supra* note 56.

68. J. M. G. Wilson & F. Jungner, Principles and Practice of Screening for Disease (1968).

69. G. Flanders et al., *Prevention of Type 1 Diabetes from Laboratory to Public Health*, 29 AUTOIMMUNITY 235 (1999); A. J. Clarke, *Newborn Screening, in* 107 Genetics, Society and Clinical Practice (P. S. Harper & A. J. Clarke eds., 1997); D. B. Bailey Jr, *Newborn Screening for Fragile X Syndrome*, 10 MENT. RETARD. DEV. DISABIL. RES. REV. 3 (2004).

70. Flanders et al., *supra* note 69; Clarke, *supra* note 69; Bailey Jr, *supra* note 69; N. A. Holtzman et al., *Effect of Informed Parental Consent on Mothers'*

Knowledge of Newborn Screening, 72 PEDIATRICS 807 (1983); L. Feuchtbaum et al., *California's Experience Implementing a Pilot Newborn Supplemental Screening Program Using Tandem Mass Spectrometry,* 117 PEDIATRICS S261 (2006); K. Atkinson et al., *A Public Health Response to Emerging Technology: Expansion of the Massachusetts Newborn Screening Program,* 116 PUBLIC HEALTH REP. 122 (2001).

B – Introduction:
Newborn Screening: Storage and Access for Research?

Ellen WRIGHT CLAYTON

Vanderbilt University, USA

The use of residual newborn blood spots for research poses an interesting array of opportunities and dilemmas. In many countries where newborn screening is performed, these spots are obtained from virtually every newborn. They represent unusually complete population-based samples, which can be tested for genetic variants and a variety of metabolic and other markers. These samples, however, are quite limited in quantity, particularly when compared with immortalized cell lines or even with blood samples typically obtained from adults. Relatively little can be learned from the blood spots themselves beyond the distribution of a measured variable among a particular population of neonates. Very little phenotypic information is contained on most birth certificates, and typically little is directly known about the "history" of most newborns. The greatest value for research occurs when children's blood spots can be linked with subsequent records of their lives regarding such matters as medical care and education.

Newborn screening typically is performed without parental permission, justified on the ground that routine testing is warranted to detect potentially treatable disorders. Parents rarely know about research using these blood spots. Consent, however, is typically required for most research. Should parental permission be required for this type of research? If so, how should it be obtained? Has too much emphasis been placed on autonomy and voluntariness at the expense of solidarity and the public good of greater knowledge? Does the public's decision to conduct this research provide a sufficient ethical foundation even in the absence of parental permission? Does the nature of the public's decision matter? For example, does a bill enacted by the legislature confer greater legitimacy than a decision by an administrative agency? What sorts of oversight are needed, particularly in light of the need for stewardship of newborn blood spots, which are small and non-renewable? What level of risk is posed by this risk, and how can it be reduced? Should results relating to a particular child ever be shared with the parents?

These and other questions raise questions at the intersections of science, research ethics, and political theory. In this section, two authors from two countries that have different political and ethical histories will provide enlightening perspectives on these questions and the dilemmas that they present, which shed light both on research involving using residual newborn blood spots and the more pervasive questions raised by the use of DNA databanks.

The Danish Newborn Screening Biobank in Practice and Research: Revised Biobank Regulations

Bent NORGAARD-PEDERSEN and David M. HOUGAARD

Statens Serum Institut, Denmark

1. INTRODUCTION

For more than two decades, all dried blood spot samples (DBSS) from the Danish Newborn Screening programme have been stored in a newborn screening biobank at - 20°C. Storage has taken place according to regulations from the Danish Ministry of Health (1993), and recently, new guidelines for the establishment and operation of biobanks in general have been made. This chapter is an update on the operation of and regulations for biobanks and their use in practice and research. There are previous overviews,[1] but recent technological developments makes an update on the new possibilities relevant. Traditional technology required at least one punch of the DBSS 3.2mm in diameter, equivalent to approximately 3µL whole blood, to determine one analyte. This reduced the number of analytes that could be determined in the limited amount of blood available in the DBSS. High throughput multi-analyte technologies such as tandem mass spectrometry and now also the Luminex® xMAP technology open new analytical possibilities for the determination of a large number of analytes that may be of interest in relation to many diseases. Moreover, DNA chip technology and the possibility of amplifying the whole genome based on DNA extracted from only a small part of the DBSS make it possible to study the relationship between multiple genetic variations and various disorders. In the past, a large number of research studies have been carried out using newborn biobank DBSS; they will be described together with ongoing and planned studies.

During the last few years, awareness has increased about existing biobanks and the establishment of new biobanks. This is due to the fact that several health care issues cannot be solved without the systematic storage of biological material.

2.   THE DANISH NEWBORN SCREENING PROGRAMME (TABLE 1)

In Denmark, it is mandatory to offer all new parents newborn screening for phenylketonuria (PKU), congenital hypothyroidism (CH) and Toxoplasmosis (Toxo).[2] The programme is carried out by analyses of DBSS taken by a heel prick 5-7 days after birth. The number of samples per year is about 65 000 and includes newborns from Denmark, Greenland and the Faeroe Islands. Before blood sampling, the parents are informed by local health professionals, through pamphlets and through the internet.[3] The biobank information focuses on its uses for (i) documentation, retesting, quality assurance and assay improvement; (II) diagnostic use later in infancy; and (iii) research.

The parents may opt out of biobank storage at the time of testing or later, either by a written letter to the department or by registering in the central "Use of Tissue Register."[4]

Several effective safety procedures are in place. The biobank samples are stored in a separate facility; they are linked to the data forms by a unique sample number only, and access to the database archive and to the freezer facility is restricted to authorized health personnel only.

The Danish newborn screening programme will be revised in 2006. Through the use of tandem mass spectrometry (MSMS), a number of inborn errors of metabolism can be identified at birth and will therefore be included in the routine programme.

3.   THE DANISH NEWBORN SCREENING REGISTER AND BIOBANK
(TABLE 2)

In 1991, a member of the Danish Parliament raised questions to the
Health Minister concerning the existence, use and legal status of the
newborn screening biobank and thereby initiated a public debate
about the biobank at the Statens Serum Institut (SSI).[5] An inspection
of the biobank facility by the Ethical Council and members of a
parliamentary committee also took place. This resulted in the biobank
at SSI being incorporated under the Public Register Act,[6] and
administrative rules were set up regarding the use of samples for
purposes other than screening. These regulations from the Danish
Ministry of Health (executive order 1993) ensured political,
administrative and legal assessments of the biobank and established a
well-defined set of regulations for its operation.[7]

The purposes of the newborn screening register and the biobank,
according to the regulations, are:

1.   Diagnosis and treatment of PKU, CH and toxoplasmosis;
2.   Control, documentation and possibility of repeated analyses if
      any of the three diseases are suspected later in childhood;
3.   Quality control and development of new screening methods
      (assay improvement);
4.   Non-individual based statistics;
5.   Specific disease testing – diagnostic use;
6.   Medico-legal uses; and,
7.   Research projects using biochemical, genetic and
      environmental markers.

In 2005, a Steering Committee for scientific use of the biobank
was set up with the main purpose of administering paragraph 7 of
these regulations. Since the blood samples contain only a limited
amount of blood, further use after routine neonatal screening must be
prioritised to ensure that enough blood is left to serve the most
important purposes, which are the following:

1st priority: analysis of the blood for the benefit of the child and
    parents;
2nd priority: the development of new methods of analysis;
3rd priority: research projects.

The initiative for establishing the Steering Committee stemmed
from a request from the Danish Medical Research Foundation after a
major grant for the establishment of a complete database for "The
Biobank as a National Research Resource." The Steering Committee
for the scientific use of the biobank is appointed every three years by
the managing director of SSI and consists of three members external
to SSI and two members working at SSI.

After approval from the Danish Data Protection Agency and the
Scientific Ethical Committee System, the Steering Committee decides
which research projects are to make use of blood samples and
information from the register and biobank. In practice, the Committee
should ensure that there is always enough blood left for each sample
to complete the necessary medical analyses in relation to the original
purpose for storage (see 1st priority).[8]

4.   THE NEW BIOBANK REGULATIONS

The new regulations implement EU-directive 95/46/EC[9] on the
protection of individuals with regard to the processing of personal
data and on the free movement of such data. A detailed account on
biobanks was published in May 2002, based on work of the task force
on the need for further legislative regulation for biobanks.[10] The task
force has defined a biobank as follows: "A Biobank is defined as a
structured collection of human biological material which is accessible
under certain criteria, and where information contained in the
biological material can be traced back to individuals." According to
the task force, a biobank can be regarded as a so-called "manual
register," subject to the *Act on Processing of Personal Data.*[11]
Consequently, the task force found that the personal data act[12] in
conjunction with the relevant legislation on health and research (*Act
on the Legal Status of Patients,*[13] *Act on a Scientific Ethical*

Committee System and the Handling of Biomedical Research Projects,[14] *Act* on central management of public health service[15] and other *Acts*) sufficiently regulates the majority of issues surrounding biobanks.

The rules of the existing *Acts* with regard to, respectively, setting up, closing, controlling and supervising biobanks and the rules on the rights of tissue donors are sufficient to secure the consideration of patients' self-determination and integrity, balanced against the considerations of research and society. However, based upon recommendations from the task force, new legislation (amendments to existing laws) has been made concerning:

1. The establishment of a central "opt out register" for the use of stored tissue ("The Register for Application of Tissue"). This register provides the option of opting out from non-treatment-related use of any biological material and also a right to destruction or conditioned right to surrender donated biobank material;
2. The applications of all research projects using biobank material should be approved by a science-ethical committee.[16]

The new regulations on biobanks are published as guidelines from the Ministry of Health 22 September 2004.[17] These guidelines are very similar to the regulations for the newborn screening biobank and register made by the Ministry of Health in 1993.[18]

The requirements are:

1. The biobank and register must be registered and accepted by the Danish Data Protection Agency with information about their purpose, operation, data-responsible authority, the person responsible for the biobank, etc.[19]
2. According to the *Act* on patients' rights.[20] This law concerns "self determination" within a clinical biobank and addresses informed consent and the right to "opt out," as well as the "destruction" or the "retrieval" of biobank material.

3. Procedures for use of biobank materials for research must always be accepted by the Scientific Ethical Committee System according to the *Act* on Scientific Committee.[21]

4. According to *Act* on Health,[22] biobank health professionals are responsible according to the general rules for healthcare personnel concerning secrecy, confidentiality etc. Complaints about biobanks can be directed to the Health Care Patients Complaints Authority, Danish National Board of Health.

In the past, these requirements have been followed by those wishing to use the newborn screening biobank for research purposes. The newly established biobank Steering Committee will ensure, as an extra safeguard, that this is also the case in the future. There have been no examples of misuse of the newborn screening biobank and register and the biobank has therefore become a leader for other biobanks.

5.   USE OF THE NEWBORN SCREENING REGISTER AND BIOBANK (TABLE 2)

First of all, the stored information is used for *diagnosis and treatment* of PKU, CH and Toxo, including control, documentation and repeated analysis if any of the diseases screened for should develop later in infancy. Retention of the original DBSS may be the only way to ascertain and document if a sample mix-up in the laboratory has taken place.

The storage is also important for *quality assurance* and *assay improvement*. Evaluation of improved screening methods for PKU, CH and Toxo cannot be carried out if case/control samples are not available. For PKU screening, tandem mass spectrometry (MSMS) was introduced after comparison with the Guthrie test; for CH, a more sensitive, precise and faster time-resolved fluorimetric sandwich assay (DELFIA) replaced the old in-house radioimmunoassay; and finally, for Toxo, a commercially available Toxo IgM assay (DELFIA) replaced a similar in-house assay. The introduction of these improved assays has been carried out without the notification of the Data Protection Agency or approval from the Scientific Ethical Committee System.

For *specific disease testing e.g. diagnostic use* in cases of unexpected morbidity or mortality during infancy, the DBSS may be used in determining the causes. Usually, request for these examinations are made by paediatricians, clinical geneticists and forensic pathologists after informed consent by the parents. Sometimes it is even possible to include examination of other siblings, if biobank DBSS are available.

For congenital infections suspected of being present at birth or after the neonatal period, the only way to make a definitive diagnosis is to analyse the DBSS for specific IgM antibodies, nucleic acid or antigens from the suspected pathogen. Examination of cytomegalovirus (CMV) infection by IgM and PCR or Toxoplasmosis by IgM may be important for identification of ethiology of hearing loss, retarded development, hydrocephalus or other abnormalities found by cerebral imaging (intracerebral calcifications).

Retrospective testing of DBSS by *biochemical tests or genetic analyses* have in the past yielded diagnostic information for a number of diseases. Such testing can have important implications for the families involved concerning the death of older siblings and/or newly diagnosed cases, as well as for genetic counselling and reproductive choices in future pregnancies. Biochemical analyses have given diagnostic information for carbohydrate deficient glycoprotein (CDG) syndrome by electrophoretic analyses of transferring,[23] peroxisomal deficiencies by gas chromatography and mass spectrometry and fatty acids oxidation disorders.[24] Genetic testing using DNA extracted from a DBSS has mainly been carried out in cases where a proband has died before genetic disease was suspected, or before a genetic analysis was done. Examples include congenital epidermolysis bullosa and genetic ion channel defects causing Long QT Syndrome.[25]

Research studies using DBSS biobank material involve mainly *retrospective screening, allele frequency* studies and studies of *etiology of a number of late onset disorders* such as Type I Diabetes, Schizophrenia, Autism, Cerebral palsy, etc. The patients are identified through medical registers from clinical departments and from unique Danish public health registers containing information on different diseases together with personal identity numbers (CPR-number). With this number, it is possible to retrieve the biobank DBSS from the respective patient cases) as well as corresponding samples from healthy persons (controls). After collecting such case/control DBSS material, the study is made anonymous and the examinations can be carried out as a so-called register type study, after approval by the Data Protection Agency, the Scientific Ethical Committee and the Biobank Steering Committee. Some studies have been carried out after written informed consent, and usually only about 1% refused to participate.[26]

Retrospective case/control samples from the biobank represent a highly efficient way to evaluate new screening techniques. Such evaluations have been carried out for cystic fibrosis using a two-tired immunoreactive trypsin - ΔF508 (IRT/ΔF508) approach,[27] for congenital toxoplasmosis using an assay for toxoplasma-specific IgM,[28] for congenital adrenal hyperplasia (CAH) using a delayed fluorescence immunoassay for 17-hydroxyprogesterone and for different inborn errors of metabolism (IEM) using quantitative analyses for amino acids and acyl carnitines by tandem mass spectrometry (MSMS). All four retrospective screening studies confirmed that screening could be started without the need for expensive prospective trials.

Toxoplasmosis was added to the newborn screening programme in January 1999, and MSMS screening will also be added to the routine programme in 2006, after having undergone several years of evaluation. CAH will probably also be introduced, whereas neonatal CF will not be implemented, since the Danish screening community favours antenatal/preconceptual carrier screening. Preliminary data based on the use of DBSS from the Danish newborn screening biobank have shown that it may be possible to screen for lysosomal disorders using Lumina multiplex technology, but further confirmation is needed.[29]

Since the DBSS biobank contains unselected material with essentially universal coverage of the entire populations of Denmark, Greenland and the Faeroe Islands since 1982, it is an optimal biobank for *allele frequency studies*. These studies can be performed as anonymous register-type studies after approval by the Scientific Ethical Committee System. The allele studies include mutations for genes for medium-chain acyl CoA dehydrogenase deficiency (MCAD),[30] apolipoprotein B-3500,[31] Factor V. Leiden,[32] hereditary haemochromatosis,[33] lutenizing hormone,[34] follicle stimulating hormone receptor,[35] transforming growth factor alpha[36] and Byler's disease.[37]

Infectious disease and vaccination epidemiology is another area where useful information can be obtained, especially when DBSS are matched with maternal samples from the same pregnancy. Such paired mother-child samples have been used to determine the sero conversion and maternal-fetal transmission rates for parvovirus B19[38] and toxoplasmosis.[39] For vaccination status, it is possible to monitor the efficiency of different programmes in the period from 1982 to 2006. Antibody monitoring (IgG) in biobank DBSS can therefore be used to monitor the mother's vaccination status over a longer period.

6.   RESEARCH AND NON-RESEARCH PROJECTS

In the new regulations on biobanks, four types of biobanks are described:

1. *Clinical biobanks used for healthcare purposes*
2. *Research biobanks used only for research purposes*
3. *Donor biobanks used for treatment of a patient or a group of patients*
4. *Biobanks with other health care purposes, e.g. stem cell biobanks, production biobanks etc.*

For practical purposes, it is only the clinical biobanks and the research biobanks which are relevant in the distinction between research and non-research projects. The ethical guidelines for the use of biobanks have been published by the Central Scientific Ethical Committee and are available in English at http://www.cvk.im.dk.

The clinical or healthcare biobanks' samples are collected and stored in relation to diagnosis and treatment. The newborn screening biobank is a good example in which storage is related to healthcare purposes. However, such biobanks related to healthcare can also be used for research projects that are closely or more distantly related to the original purpose of the biobank. Therefore, the person responsible for the biobank should ensure—before biobank material is used for research purposes—both that the donors have not registered in the central use of tissue register and that the project has been approved by the Data Protection Agency and the Scientific Ethical Committee System. Generally, informed consent should be requested. For practical purposes, it is possible to obtain permission without consent according to the so-called health care exemption. However, this exemption should always be approved by the Ethical Committee.

Research biobanks are collected and stored for research purposes only and therefore always include informed consent. It is recommended, however, that the initial consent be formulated for a wide range of uses, so that it is not necessary to have new consent for all projects. Again, this exemption should be accepted by the Ethical Committee. A good example of a large research biobank is the Danish National Birth Cohort ("Better health for mother and child") where blood samples have been taken twice during pregnancy from the mother and cord blood is taken at birth.[40] This biobank includes 100 000 pregnant women and corresponding newborns. The study also includes a number of interviews, and it is expected that a large number of gene-environmental hypotheses can be based on case-control studies from this national cohort. Several of the studies may also benefit from the newborn DBSS Biobank.

Recently, a publication on the Danish PKU biobank has been made by representatives from the Copenhagen University Law Faculty.[41] The inspection/interview dealt with research/non-research projects based on the newborn screening biobank and research in law and in practise. The question was how we used the new rules and guidelines in the definition of research versus non-research. Generally, three pragmatic distinctions have been used in the past: namely, *close to or distant* from the original purpose of the biobank, *internal/external* involvement and the *prospective/retrospective distinction*. Generally, they found that in the past, we have lived up to the new guidelines although a few "projects" were labelled "research" by the authors though we called them healthcare developmental projects. The lessons from this inspection may be that, in the future, all projects using biobank material will be called "research" projects although this may, in practise, be quite bureaucratic and time-consuming.

In conclusion, the Danish newborn screening biobank (PKU-biobank) is regulated by specific legislation with oversight. Parents are informed at specimen collection about the biobank and its use, partly verbally and partly through pamphlets and a website[42] Information focuses on the use of specimens. Parents may choose to opt out from storage after routine screening has been performed or later, by using the central national "Use of Tissue Register" for Denmark.

The storage of the newborn DBSS includes secured privacy protection with controlled access. The link between data/specimens is labelled only by a unique number. In order to identify a specific DBSS, this number should be found in the computer database. The samples are stored in a large freezer at - 20°C with access for authorized personnel only.

The newborn screening biobank is indeed a National Research Resource which is used for a number of research projects to benefit the child, the parents, society and future generations.

ACKNOWLEDGEMENTS

The Danish National Newborn Screening Biobank has obtained financial support from the Danish Medical Research Foundation grant No. 22-03-0458.

Table 1: Routine newborn screening for PKU (1975), CH (1978), and Toxoplasmosis (1999) in Denmark, Greenland and the Faroe Islands

- Mandatory to offer screening
- Informed dissent (+/- screening, - storage)
- PKU - biobank since 1982
- Cost benefit ratio 1:28
- Accreditation: Iso 17025 by Danak 1998 and onwards
- New guidelines for newborn screening 2006

Table 2: The Danish Newborn Screening Register and Biobank

The stored information and biobank is used for:

1. Diagnosis and treatment of PKU, CH and Toxoplasmosis
2. Control, documentation and repeated analyses
3. Quality assurance and assay improvement
4. Non-individual based statistics
5. Specific disease testing – diagnostic use
6. Medico-legal use
7. Research projects using biochemical, genetic and environmental markers

References

1. B. Nørgaard-Pedersen, *Use of Stored Samples from the Danish PKU Register, in* 303 Human DNA: Law and Policy (B. M. Knoppers & C. M. Laberge eds., 1997); B. Nørgaard-Pedersen, *The Danish PKU Register and Biobank, in* 59 Proceedings from the Workshop on Human Biobanks: Ethical and Social Issues (vol. 9, Nord Biotechnol. Series) (1997); B. Nørgaard-Pedersen & H. Simonsen, *Biological Specimen Banks in Neonatal Screening,* 432 Acta Pædiatr. (supplement) 106 (1999).
2. H. Simonsen et al., *Neonatal Screening i Danmark. Status og Fremtidsperspektiver,* 160 Ugeskr. Laeg. 5777 (1998).
3. Statens Serum Institut, http://www.ssi.dk (gravide/nyfødte).
4. Sundhedsstyrelsen, http://www.Sundhedsstyrelsen.dk/vaev.
5. G. Almind et al., Health Science Information Banks: Biobanks (1996).
6. *Public Register Act,* available in An Account on Biobanks (Summary in English), 1414 (2002).
7. *Regulations for the PKU Registry,* Ministry of Health Executive Order, Jan. 14, 1993, Copenhagen.

8. Statens Serum Institut, 2004: 1-3.
9. Council Directive 95/46, 1995 O. J. (L 281) 31.
10. An Account on Biobanks (Summary in English), 1414 (2002), at 245.
11. *Act on Processing of Personal Data,* No. 429, May 31, 2000 (Dan.).
12. *Id.*
13. *Act on the Legal Status of Patients,* No. 482, July 1, 1998 (Dan.).
14. *Act on a Scientific Ethical Committee System and the Handling of Biomedical Research Projects,* No. 402, May 28, 2003 (Dan.).
15. No. 141, March 5, 2001 (Dan.).
16. Anonymous, Guidelines about Notification etc. of a Biomedical Research Project to the Committee System on Biomedical Research Ethics (2004), http://www.cvk.im.dk/visArtikel.asp?artikelID=1642.
17. *Vejledning om biobanker inden for sundhedsområdet: Patientrettigheder og mundighedskrav* [Guidelines for Biobanks in the Health Care Area: Patient Rights and Authority Claims], 1 SK. T., J. NR. (2004)1660-24.
18. *Regulations for the PKU Registry, supra* note 7.
19. *Act on Processing of Personal Data, supra* note 11.
20. *Act on Patients' Rights,* available in An Account on Biobanks (Summary in English), 1414 (2002).
21. *Act on Scientific Committee,* available in An Account on Biobanks (Summary in English), 1414 (2002).
22. *Act on Health,* available in An Account on Biobanks (Summary in English), 1414 (2002).
23. M. B. Petersen et al., *Early Manifestations of the Carbohydrate-Deficient Glycoprotein Syndrome,* 122 J. PEDIATR 66 (1993).
24. C. Jacobs et al., *Diagnosis of Zellweger Syndrome by Analysis of Very Long-Chain Fatty Acids in Stored Blood Spots Collected at Neonatal Screening,* 16 J. INHERIT. METAB. DIS. 63 (1993).
25. L. A. Larsen et al., *Recessive Romano-Ward Syndrome Associated with Compound Heterozygosity for Two Mutations in the KVLQT1 Gene,* 7 EUR. J. HUM. GENET. 724 (1999); M. Christiansen et al., *Mutations in the HERG K+-ion Channel: A Novel Link Between Long QT Syndrome and Sudden Infant Death Syndrome,* 95 AM. J. CARDIOL. 433 (2005).
26. K. Christensen et al., Oral Clefts, *Transforming Growth Factor Alpha Gene Variants, and Maternal Smoking: A Population-Based Case-Control Study in Denmark, 1991-1994,* 149 AM. J. EPIDEMIOL. 248 (1999).
27. B. Nørgaard-Pedersen et al., *Immunoreactive Trypsin and a Comparison of Two ΔF508 Mutation Analyses in Newborn Screening for Cystic Fibrosis: An Anonymous Pilot Study in Denmark* 2 SCREENING 1 (1999).
28. M. Lebech et al., *Feasibility of Neonatal Screening for Toxoplasma Infection in the Absence of Prenatal Treatment,* 353 LANCET 1834 (1999).
29. P. J. Meikle et al., *Newborn Screening for Lysosomal Storage Disorders,* MOL. GEN. METAB. (forthcoming 2006).
30. J. B. Lundemose et al., *The Frequency of a Disease-Causing Point-Mutation in the Gene Coding for Medium-Chain Acyl-CoA Dehydrogenase (MCAD) in Sudden Infant Death Syndrome (SIDS),* 82 ACTA PAEDIATR. 544 (1993).
31. P. S. Hansen et al., *Incidence of the Apolipoprotein B-3500 Mutation in Denmark,* 230 CLIN. CHIM. ACTA 101 (1994).

32. T. B. Larsen et al., *The Arg506Gln Mutation (FV Leiden) among a Cohort of 4188 Unselected Danish Newborns*, 89 THROMB. RES. 211 (1998).

33. A. T. Merryweather-Clarke et al., *A Retrospective Anonymous Pilot Study in Screening Newborns for HFE Mutations in Scandinavian Populations*, 13 HUM. MUTAT. 154 (1999).

34. C. Nilsson et al., *Determination of a Common Genetic Variant of Luteinizing Hormone Using DNA Hybridization and Immunoassays*, 49 CLIN. ENDOCRINOL. (OXF.) 369 (1998).

35. M. Jiang et al., *The Frequency of an Inactivating Point Mutation (566C→T) of the Human Follicle-Stimulating Hormone Receptor Gene in Four Populations Using Allele-Specific Hybridization and Time-Resolved Fluorometry*, 83 J. CLIN. ENDOCRINOL. METAB. 4338 (1998).

36. Christensen, *supra* note 26.

37. H. Eiberg et al., *Cholestasis Familiaris Groenlandica/ Byler-Like Disease in Greanland – A Population Study*, Proceedings of the 12th International Congress on Circumpolar Health, 63 INT. J. CIRCUMPOLAR HEALTH (supplement 2) 189 (2004).

38. A. K. Valeur-Jensen et al., *Risk Factors for Parvovirus B19 Infection in Pregnancy*, 281 J. A. M. A. 1099 (1999).

39. M. Lebech & E. Petersen, *Neonatal Screening for Congenital Toxoplasmosis in Denmark: Presentation of the Design of a Prospective Study*, 84 SCAND. J. INFECT. DIS. (supplement) 75 (1992).

40. J. Olsen et al., *The Danish National Birth Cohort – Its Background, Structure and Aim*, 29 SCAND. J. PUBLIC HEALTH 300 (2001).

41. M. Hartlev & U. Lind, *Use of Blood Samples from the Danish PKU-Biobank – A Study of the Conceptualization of Research in Law and in Action*, Eureca Workshop, Tenerife 2005 (forthcoming 2006).

42. Statens Serum Institut, http://www.ssi.dk.

Research and Public Health Surveillance Using Newborn Bloodspots in Canada

Denise AVARD

Centre de recherche en droit public, Université de Montréal, Montréal

1. INTRODUCTION

Newborn screening programmes are highly effective public health programmes that exist on all continents.[1] The essential goal of newborn screening is to test early for disorders which can be prevented or for disorders whose severity can be reduced by timely intervention. These criteria are best exemplified by phenylketonuria (PKU), for which early screening and treatment has been very effective in preventing neurological abnormalities.[2] Considering the approximately 328 802 births per year in Canada,[3] with each child being tested for at least two and up to 30 genetic and non-genetic diseases (depending on their province of birth), newborn screening is the most extensive genetic screening programme in this country.[4]

While newborn screening programmes provide considerable health benefits for children, the blood samples collected as part of routine newborn screening programmes have acquired a new and significant value as a result of the development of genomics.[5] Generally, after newborn screening is completed, the dried blood spots are stored in public health laboratories, for variable periods of time, in order to permit confirmatory diagnosis, re-testing if needed, and quality control.[6] The storage of dried blood spots for these purposes is not particularly controversial since these uses are related to the primary purpose of the initial collection: namely, screening for various disorders. But dried blood spots may also be used for purposes unrelated to the initial collection. These blood samples are considered valuable resources for health research and possible tools to help answer important public health questions.[7]

Dried blood spots are especially useful for research purposes because they are collected from all newborns and therefore represent an unbiased sample as well as a unique population-based specimen of tissues. Using newborn dried blood samples for medical, clinical and public health research could bring great benefits, including the development of effective public health interventions and improvements in the health of babies and their families. Indeed, newborn blood spots are used to validate new screening tools and to develop new testing methods. In epidemiological studies, they are used to analyze and establish population frequencies of gene variants[8] for cystic fibrosis[9] and for Sudden Infant Death Syndrome.[10] Studies are underway to identify genetic risk factors associated with susceptibility to meningococcal and pneumococcal infections by using blood spot specimens originally collected for state-based newborn screening programmes.[11]

From a public health standpoint, the use of newborn blood spots is invaluable for public health surveillance, to identify disease incidence and prevalence, and to identify who in a population is at risk for disease. For example, newborn bloodspots have been used as a convenient and inexpensive data source to carry out health surveillance such as anonymous surveillance studies for HIV prevalence and hepatitis C virus surveillance.[12] Collecting and using newborn blood spots and other health-related information can be helpful in identifying health problems, developing solutions and facilitating appropriate interventions,[13] thereby improving the health of the populations. For example, a project in Atlanta, Georgia (USA) linked blood spot data from their newborn screening programme to data from the Developmental Disabilities Surveillance programme and the special education programme in order to follow-up on the long-term developmental and neurological outcomes of children who screened positive a birth for a metabolic or endocrine disorder. This information helped with the planning for the special health, education and newborn screening services in the region.[14] As another example, in the United Kingdom, anonymous testing of newborn blood spots for maternal anti-HIV antibody is carried out to determine whether or not antenatal HIV testing is being successfully implemented.[15]

However, there is concern over the desirability of storing newborn bloodspots, especially in the absence of clear guidance, because storing dried blood spots raises a number of socio-ethical and public policy issues.[16] Genetic data can be stored indefinitely and has the potential to reveal an infinite amount of personal information both about an individual and about the individual's siblings, parents and relatives. Moreover, the information may be of interest to third parties such as insurers, employers and schools, in addition to its useful for forensic purposes or paternity testing.[17] Some observers fear that the ethical issues associated with the use of newborn blood spots are of even greater concern because of the mandatory nature of the newborn screening programmes[18] and because of the special vulnerability and rights of children. The unique DNA information extracted from newborn blood spots raises concerns about individual privacy and familial privacy. In addition, there are fears that the use of stored DBS may lead to discrimination and even, potentially, to harm to individuals and groups. There is also concern that DBS could be used for non-medical purposes such as crime investigation or paternity testing.

As a result, there is concern over the desirability of storing newborn bloodspots when there are a lack of safeguards that protect privacy, ensure confidentiality and prevent the potential harms associated with storing and accessing the personal information linked to the newborn blood spots. Many jurisdictions around the world are grappling with the need to establish formal policies for the storage of data collected from newborn screening programmes, particularly Denmark, Britain, Australia and the United States.[19] A number of international organizations and regional bodies recognize the need to establish formal policies for the storage of data collected from newborn screening programmes.[20]

Canada has not yet taken formal steps to address these concerns about storage and use of newborn bloodspots for research. A research initiative was therefore undertaken under the leadership of D. Avard;[21] it involved co–investigators from several provinces and focused on the development of ethically grounded policy research for use by directors of newborn screening programmes and government policymakers. The goals of the research were to:

- Review the policies and practices for storing newborn blood spots;
- Survey screening laboratory policies and practices for storing newborn bloodspots in Canada;
- Build a Canadian Multidisciplinary Newborn Task Force; and,
- Host a consultative workshop addressing newborn screening programme practices regarding storage.

Newborn screening in Canada is administered and governed at the provincial level. Current practices of storing newborn blood spots reveal policy differences between provinces and, more generally, illustrate that usually no formal policies addressing the storage of newborn blood spots exist. The survey of provincial Laboratory Directors revealed a lack of written policies and procedures in several provincial newborn programmes in relation to the storage and potential future uses of stored bloodspots. Indeed, the length of storage varied significantly from province to province, ranging from one to over twenty-one years. A more in-depth analysis of the survey responses provided by the Laboratory Directors is reported in the submitted paper.[22]

This chapter will first summarize the issues, concerns and ideas that emerged during a consultative workshop undertaken as part of the investigation into of the development of policy research related to newborn screening in Canada. Secondly, by way of example, it will consider the role of Quebec's *Public Health Act* in facilitating public health research and surveillance. Finally, the chapter will conclude with a discussion about the need to promote participatory models that engage both the public and decision makers in the formulation of policies to address concerns about the storage of newborn blood spots for research purposes.

2.   STORAGE AND USE OF NEWBORN BLOODSPOTS FOR RESEARCH:
SOCIO-ETHICAL ISSUES

The goal of the consultative workshop was to engage stakeholders and to assess their views on the socio-ethical issues surrounding the storage and use of newborn bloodspot cards for research. An important venue to address the issues of storage of newborn bloodspots and to establish links with key stakeholders was at the annual 2005 meeting of the Garrod Society, where we hosted an event to achieve these objectives.[23] As evidenced in research literature, the concept of partnership and a multidisciplinary approach has been increasingly important[24] and was integral to our vision. At this event, we brought together a multidisciplinary working group of 30 stakeholders representing a cross-section of the community, including:

- representatives from provincial newborn screening programmes;
- representatives from various governmental and non-governmental organizations;
- researchers in the fields of law, biomedical ethics and sociology, and an expert in evidence-based medicine;
- research ethics committees; and
- advocacy groups.

These stakeholders were asked to consider three central themes: the secondary use of NBS for research, consent and secondary use, and arrangements to keep information private. Below are selected highlights of the issues that emerged during the workshop.

2.1   *Secondary Uses of Newborn Bloodspots for Research*

Overall, it was agreed that not all secondary research is the same: there can be appropriate and inappropriate secondary uses of newborn dried blood spots. Still, there was no consensus on where to draw the line and how best to proceed. All participants concurred, however, that if the samples are to be used for secondary purposes, the secondary use should not interfere in any way with the original mandate of the newborn screening programme.

The discussion focussed primarily on the secondary use of DBS for research purposes. There was a general if not unanimous sentiment that use for research is permissible because research projects generally receive Research Ethics Board (REB) review, and inappropriate projects will presumably be denied. Some individuals felt that existing guidelines used by REBs provide a sufficient safeguard. The Tri-Council Policy Statement, for example, provides that if it is impossible to identify individuals, then researchers should be allowed to use that database (article 3.3).[25]

2.2   Consent and Secondary Use

Newborn screening programmes in Canada are mandatory. Consent from parents before carrying out the newborn screening tests is implicit because it is considered a standard mandatory health procedure.[26] The well established model of implied consent for primary uses was not in question at the workshop; rather, the discussion centred on the more contentious issue of what quality of consent should be required for secondary uses of stored samples. There was no consensus as to whether informed consent, either of a general or specific nature, should be obtained before newborn blood spots may be used in research.

Information on newborn screening is generally provided to parents in written form as part of a pre-natal package and/or at the time of birth. A concern of the workshop participants related to whether information given to parents would be read and understood. There was consensus about the importance of informing parents generally regarding the screening programme, regardless of whether consent is implied or explicit, and of clarifying both the length of time the samples will be stored and any secondary uses that might occur, if applicable. However, it was also felt that if explicit informed consent were to become a requirement for newborn screening due to the use of samples for various secondary purposes, participation rates in screening programmes on the whole may be negatively affected. If this were the case, then the primary purpose of the screening would be impaired. This could be seen as a justification for not requiring explicit consent and relying instead on anonymization and ethics board approval: at the end of the day, the health benefits of newborn screening are too valuable to be jeopardized. Many of the participants felt that the current ethical review process is sufficient and that explicit consent is not required if samples are anonymized. However, from a healthcare perspective, there exist some arguments against anonymization because it may entail its own ethical problems if, for example, the research detects something that is of relevance to an individual's health: since the origin of an anonymized sample cannot be traced, the patient could never be informed of the findings.

Finally, the fact that consent is implicit at the time of the newborn screening test was cited as a factor that supports a shorter storage time. Moreover, it was noted that the length of storage and the types of uses envisioned will determine, to a great extent, the nature of consent required. Many felt that the possibility of using newborn blood spot for future, unspecified research projects underscores the need for a simple consent procedure for storage.

2.3   *Protection of Privacy*

One of the major concerns regarding storage of NBS is the protection of genetic information found on the newborn bloodspot and the ability to retrace the bloodspot to the donor. Stored, identifiable blood samples need to be properly safeguarded to ensure the donor's privacy is respected. If identifiers are required for a study, or if contact with patients or families is needed, some participants advocated for a number of steps such as no release of NBS or data without prior approval of the protocol by an ethical review board, including parental consent, terms of release, and confidentiality protection.

Although significant benefits may be gained from the storage of newborn blood spots for genetic testing, apprehension existed about the possible misuse of these samples because misuse could lead to discrimination, psychological harm, identification or incorrect assignment of paternity and potential social injustices.

In summary, the socio-ethical issues described by the workshop participants were similar to many of the issued raised in the literature.[27] In light of the growing interest in using newborn blood spots and the lack of clear ethical guidelines about the storage of dried blood spots, many health service providers, policymakers and researchers in Canada are grappling with the need to address proactively the best approaches related to these issues: the justifications for storage and potential secondary uses, data sharing issues, confidentiality, security and privacy issues, informed consent, and the information to be imparted to parents.

3.   NEWBORN BLOOD SPOTS AND PUBLIC HEALTH RESEARCH

Public health research distinguishes itself from medical research by attempting to answer questions related to the health of populations rather than the health of individuals and by addressing health promotion and prevention rather than cures.[28] Identifying the health of the population and tracking patterns of morbidity and mortality through surveillance are integral strategies of public health.[29] There are many potential benefits of public health surveillance. For example, surveillance helps governments, health planners, policymakers, and epidemiologists to develop policies and legislation, as well as make informed decisions regarding health programmes, to identify factors that cause certain diseases, to launch information campaigns so people can take appropriate action, to reduce the risk of public health crises, and to establish research priorities and generate research hypotheses.[30]

Surveillance in a public health context is formally defined as *"... the ongoing, systematic use of routinely collected health data to guide public health action in a timely fashion."*[31] However, personal information collected during the surveillance process for the benefit of the population under review can also be a source of harm, either intentionally or unintentionally. Hence, the use of surveillance data must be balanced against individual rights to privacy and, thus, the data must be well guarded. Personal health information is considered the most private and sensitive information, and only the minimum information necessary for a stated purpose may be collected.[32]

By way of illustration, this chapter examines what is happening in Quebec by focusing on Quebec's *Public Health Act*. The scope of the Act is based on a broad definition of public health and includes: 1) ongoing surveillance of the population's health status; 2) the promotion of health and well being; 3) the prevention of diseases, injuries and psychosocial problems that can have an impact on population health, and 4) health protection.[33]

While public health is concerned with all these objectives, determining what is understood by surveillance in the Quebec *Public Health Act* is a striking example of an approach that attempts to balance protection and health promotion while recognising the importance of the determinants of health and that the health of the population may be remarkably different for different communities.

Theoretically, public health is concerned with all these objectives. However, traditional public health practice has followed primarily the infectious disease model, which tracks and manages communicable diseases when they arise in the population. Current thinking has relegated this approach to the past because focusing exclusively on infectious diseases limits the potential positive impact of public health authorities on population health. As demonstrated in the *Public Health Act*, Quebec has taken a unique approach to the issues of population health and health promotion through its definition of the term "surveillance," through its definition of the *Act's* mandate, and by developing innovative protection mechanisms. By expanding the realm of public health to include the study of determinants of health, such as social conditions, individual behaviours and lifestyle, etc., the Quebec model attempts to achieve a greater positive impact on public health while maintaining mechanisms to protect individual public interests. Quebec thus provides a striking example of an approach that aims to balance health protection and health promotion while recognizing the scientific and social importance of determinants of health among the whole population and among different communities with the general population.

3.1   *Surveillance and Monitoring*

Quebec's *Public Health Act* divides the traditional surveillance function into two areas of activity: surveillance and health monitoring. Surveillance of the population health includes *"the continuous collection and analysis of information about the population in general regarding elements such as demographic, socio-cultural and socio-economic conditions; physical environment; lifestyle, risk factors and prevention habits; general health status; physical and mental health; and variable health care services."*[34] This data is typically used to inform decision makers and to help them make better decisions, develop policies, plan and evaluate programmes, and implement services. Surveillance data is also useful for informing the public. Public health monitoring, on the other hand, *"involves the ongoing collection of data, it targets persons who are at risk of contracting, or are affected by, an infectious disease or other threat to health. The collection of data is also more direct than in the case of surveillance and stems from the authority that public health actors have to intervene in order to protect public health."*[35] Mandatory reporting of infectious diseases (e.g. tuberculosis) and some sexually transmitted diseases (e.g. gonorrhea) to the provincial health department constitutes a public health monitoring activity and, for the purposes of the *Act*, is not considered surveillance. Mandatory screening is also a central feature of public health monitoring and, generally, such screening occurs because there is an intervention that will be helpful to the individuals involved. Thus, screening of newborns for PKU would be considered a public health monitoring activity in Quebec.

3.2   Mandate

We have seen that the Quebec Legislation authorizes the Minister of Health and Social Services and Public Health Directors to conduct surveillance activities. They are expected to develop a plan for the surveillance of the health status of the populations they serve. The *Act* authorizes the use, for surveillance, of information initially collected for other purposes. For example the Minister and the public health authorities can require for example that Directors of newborn screening laboratories disclose information they have in their possession. Thus, stored NBS from the provincial newborn screening laboratory may be requested, and the information should be sent to the health department.[36]

The Minister and the Directors of Public Health also have the authority to link information from various databases. Hence, newborn blood spots information could be theoretically linked to other survey information on health and/or social issues.

3.3   Mechanisms to Protect

The Public Health Act has several practices in place to protect information from improper use. For example, there must be a surveillance plan to describe the purpose and object of the surveillance, the type of information to be collected (individual and non-personal), the proposed sources of information, and the investigative approach needed to exercise the surveillance activities.[37]

In addition to being subject to Quebec's privacy legislation, surveillance plans and surveys submitted by the Minister or the pubic health directors must be submitted to the *Comité d'Ethique de Santé Publique (*CESP*)*.[38] The mandate of this committee is to evaluate all surveillance plans and proposed surveys and to give opinions to the Minister. In practice, the committee is not a decision-making body but a community barometer that provides moral authority to authorities. An important step for the committee was to promote a broad range of perspectives that reflect the views of both community members and public health professionals.

4.   SUMMARY AND CONCLUSIONS

There are many concerns about the secondary use, storage and destruction of newborn blood spots. While using newborn dried blood samples for medical, clinical and public health purposes has many advantages, both for the child and for public health research or other medical research, there are concerns that researchers increasingly rely on and demand newborn screening specimens for research that may not be a direct benefit for the child, such as genetic studies to understand the genetic contribution of multifactorial chronic conditions, or for public health surveillance.

However, this secondary use raises a number of socio-ethical concerns. There is no parental informed consent, in the strictest sense, for using these newborn blood spots for purposes other than for the testing of treatable conditions such as PKU. In addition to concerns for respect of autonomy and integrity, there are also unique privacy issues associated with the use of genetic information from newborn blood spots because the information is immutable, predictive, familial and even communal, and the information about an infant's genetic make-up may be of interest to third parties such as insurers, employers and schools, as well as for forensic purposes or paternity testing. Another concern is the wide variation in practice standards for the storage and use of newborn blood spots across Canada and internationally. One final concern is the lack of awareness by the public that these newborn blood spots are being stored and used for other purposes that those initially planned.

Handling such complex issues requires a consideration of the perspectives and values of a wide range of stakeholders who may be impacted by these issues, including families of affected and non-affected infants, healthcare professionals and the broader public. Increasingly in health policy, development practitioners, decision makers, policymakers and the public are being urged to work together.[39]

In recent years, the concept of participatory models of policy development have received a great deal of attention.[40] From the outset, Quebec's *Public Health Act* has recognized the importance of creating a forum to discuss the ethical aspects of proposed public health surveillance strategies. In the UK, research by members of the Social Science Research Unit at the Institute of Education is addressing the storage and use of newborn blood spots issue using a public consultation process to ask people their views about the uses of newborn blood cards.[41] Also, Australia has been proactive in clarifying the issues relating to storage and access to the newborn screening cards. The report raises concerns about parent's lack of knowledge and recommends that parents be able to consent separately to newborn screening cards being used for research purposes.[42]

The government of Quebec approach is consistent with others and bringing together leaders in the newborn screening community in a consultative workshop to identify pivotal socio-ethical issues regarding the storage of newborn blood spots was an important step in a strategy to build policies and links with key stakeholders. In light of the lack of uniformity in provincial approaches to dried blood spot storage, uses and access, and the amount of information given to parents about storage, and considering the particularly sensitive nature of the information that can be derived from dried blood spots, there is a need for heightened transparency and clear recommendations concerning the criteria for storage, the length of storage, and permissible secondary uses of dried blood spots. The sooner this clarity emerges, the better for the health and security of all Canadians.

ACKNOWLEDGEMENTS

The author thanks the following people for their assistance with the project: Hilary Vallance, Cheryl Greenberg, Claude Laberge, Beth Potter, Margo Plant, Linda Kharaboyan and the Canadian Institute of Health Research for funding the project.

REFERENCES

1. R. Harris & M. Reid, *Medical Genetic Services in 31 Countries: An Overview*, 5 EUR. J. HUM. GENET. (supplement) 3 (1997); J. J. Stoddard & P. M. Farrell, *State-to-State Variations in Newborn Screening Policies*, 151 ARCH. PEDIATR. ADOLESC. MED. 561 (1997).

2. M. J. Thomason et al., *A Systematic Review of Evidence for the Appropriateness of Neonatal Screening Programmes for Inborn Errors of Metabolism*, 20 J. PUBLIC HEALTH MED. 331 (1998).

3. Anonymous, *Births*, THE DAILY, Apr. 19, 2004, http://www.statcan.ca/Daily/ Francais/040419/q040419.pdf.

4. W. B. Hanley, *Newborn Screening in Canada – Are we Out of Step?*, 10 PAEDIATRICS AND CHILD HEALTH 203 (2005).

5. C. Laberge et al., *Newborn Screening, Banking, and Consent*, 2 GENEDIT 1 (2004).

6. Laberge, *supra* note 5; D. Avard et al., *Variability in the Storage and Use of Newborn Dried Bloodspots in Canada: Is it Time for National Standards*, (forthcoming 2006).

7. J. E. McEwen & P. R. Reilly, *Stored Guthrie Cards as DNA "Banks"*, 55 AM. J. HUM. GENET. 196 (1994).

8. The New York State Task Force on Life and the Law, Genetic Testing and Screening in the Age of Genomic Medicine 406-411 (2000).

9. B. Norgaard-Pederson, *Use of Stored Samples from the Danish PKU Register*, *in* 303 Human DNA: Law and Policy, International and Comparative Perspectives (B. M. Knoppers ed., 1997).

10. D. Abramson, *Passing the Test: New York's Newborn HIV Testing Policy, 1987-1997*, *in* 313 Reducing the Odds – Preventing Perinatal Transmission of HIV in the United States (M. A. Stoto et al. eds., 1999).

11. S. Zimmer, *State Uses of Residual Blood Spots for Public Health: Evaluating Genetic Susceptibility to Meningcococcal and Pneumocococcal Disease Using Dried Blood Spots* (2006), http://www.aphl.org/conferences/genetic_testing _symposium.

12. C. Hankins, *Research on Surveillance: Anonymous Seroprevalence Studies-Ethical and Epidemiological Considerations in Public Health Planning*, *in* 87 Genetic Screening. From Newborns to DNA Typing (B. M. Knoppers & C. M. Laberge eds., 1990); A. C. Toledo Jr. et al., *Dried Blood Spots as a Practical*

and Inexpensive Source for Human Immunodeficiency Virus and Hepatitis C Virus Surveillance, 100 MEM. INST. OSWALDO CRUZ 365 (2005).

13. L. W. Chambers et al., *Health Surveillance: An Essential Tool to Protect and Promote the Health of the Public,* 97 CAN. J. PUBLIC HEATH 11 (2006).

14. B. K. Van Naarden et al., *Long-Term Developmental Outcomes of Children Identified through a Newborn Screening Program with a Metabolic or Endocrine Disorder: A Population-Based Approach,* 143 J. PEDIATR. 236 (2003).

15. D. Goldberg & L. Logan, *Unlinked Anonymous Testing Indicates Antenatal HIV Testing in England and Scotland Is Being Successfully Implemented,* 10 EURO. SURVEILL. E050519.4 (2005), available at http://www.eurosurveillance.org/ew/ 2005/050519.asp#4.

16. D. Avard & B. M. Knoppers, *Screening and Children,* 2 CAN. J. POLICY RES. 46 (2001); E. W. Clayton, *What Should Be the Role of Public Health in Newborn Screening and Prenatal Diagnosis?* 16 AM. J. PREV. MED. 111 (1999); M. Z. Pelias, *Research in Human Genetics: The Tension between Doing no Harm and Personal Autonomy,* 67 CLIN. GENET. 1 (2005).

17. K. Elkin & D. G. Jones, *Guthrie Cards: Legal and Ethical Issues,* 1 N. Z. BIOETH. J. 22 (2000).

18. The New York State Task Force on Life and the Law, *supra* note 8.

19. L. Kharaboyan et al., *Storing Newborn Blood Spots: Modern Controversies,* 32 J. LAW MED. ETHICS 741 (2004).

20. Laberge et al., *supra* note 5.

21. Avard et al., *supra* note 6.

22. *Id.*

23. Garrod Symposium, Ottawa, 2005, http://www.garrod.ca/.

24. E. H. Hiller et al., *Public Participation in Medical Policy-Making and the Status of Consumer Autonomy: The Example of Newborn Screening Program in the United States,* 87 AM. J. PUBLIC HEALTH 1280 (1997).

25. Medical Research Council of Canada, Natural Sciences and Engineering Research Council of Canada, Social Sciences and Humanities Research Council of Canada, Tri-Council Policy Statement – Ethical Conduct for Research Involving Humans (1998).

26. Avard & Knoppers, *supra* note 16.

27. Kharaboyan et al., *supra* note 19.

28. M. J. Khoury et al., Genetics and Public Health in the 21st Century. Using Genetic Information to Improve Health and Prevent Disease (2000).

29. B. von Tigerstrom et al., *Legal Regulation of Cancer Surveillance: Canadian and International Perspectives,* 8 HEALTH LAW J. 1 (2000).

30. Chambers et al., *supra* note 13.

31. J. Hockin, *Surveillance, in* 1171 Encyclopedia of Public Health (L. Breslow ed., 2001).

32. Personal Information Protection and Electronic Documents Act, 2000 S.C., ch. 5 (Can.).

33. M. Lacroix, *Quebec's Public Health Ethics Committee: A Model for the Public Health Agency of Canada,* 43 ALBERTA LAW REV. 511 (2005).

34. *Id.* at 516.

35. *Id.* at 517.

36. Public Health Act, R.S.Q., ch. S-2.2, section 55 (Que.).

37. Lacroix, *supra* note 33.

38. Public Health Act, sections 20, 21, 43.

39. J. Lomas, *Connecting Research and Policy*, 1 CAN. J. POLICY RES. 140 (2000); J. Lomas et al., Conceptualizing and Combining Evidence for Health System Guidance (2005), http://www.chsrf.ca/other_documents/pdf/evidence_e.pdf.

40. A. J. Culyer, *Involving Stakeholders in Health Care Decisions: the Experience of the National Institute for Health and Clinical Excellence (NICE) in England and Wales*, 8 HEALTH QUARTERLY 56 (2005).

41. S. Oliver et al., The Storage and Use of Newborn Babies' Blood Spot Cards: A Public Consultation 1 (2005).

42. C. Nader, *Baby Tests Need Parent Input: Report,* THE AGE, Aug. 16, 2006, http://www.theage.com.au/news/childrens-health/baby-tests-need-parent-input-report/2006/08/16/1155407859208.html.

Implementation of Population Screening

Jane HALLIDAY

Public Health Genetic, Murdoch Childrens Research Institute, Victoria, Australia

Implementation of population screening is the subject of many papers and reports by experts; this chapter is an accessible summary of issues related to the implementation of population screening.

1.   TESTING VERSUS SCREENING

There is an important difference between traditional genetic testing and population genetic screening.[1] Genetic testing is offered in **high risk situations**, along with genetic counselling, as best practice healthcare when there is a strong family history and/or when a person already has symptoms of a genetic condition. Populations at **increased risk** can have opportunistic testing when they routinely come into contact with the health care system or as part of a specially designed convenience screening programme. Classic examples are screening of Ashkenazi Jews for Tay Sach disease, French Canadians for familial hypercholesterolemia and many racial groups for beta thalassaemia. And then there is non-targeted, systematic population screening for those at **average** risk.

Screening programmes for asymptomatic persons at average risk are shown here in the order they have appeared in various healthcare systems around the world.

1) Newborn screening
2) Antenatal screening of pregnant women. Both are captive populations.
3) Other adult, consenting, average-risk populations for
    a) carrier status (e.g. Cystic Fibrosis, CF) or
    b) predictive testing for relatively common chronic diseases with a single gene component. (e.g. Hereditary Haemochromatosis, HH)
4) On the horizon is screening for susceptibilities to complex common disease with genetic markers (mutations or polymorphisms) once clearly identified.

2.   A WEEK IN THE LIFE OF DEE & AIDAN CHIP

To show how population screening is invariably connected to genetic testing and the complexities that can arise for individuals having been screened, consider the example of Dee & Aiden Chip, a young couple in 2007, somewhere in the developed world, Dee has had population screening for CF by ordering a test over the internet and mailing her cheek brush swab to a commercial lab. Aidan has also been population screened (convenience screened) for HH at his workplace.[2] They discover that they are both carriers. What do they do with this genetic information they have gained from such population screens, especially as Dee is pregnant?

In countries with universal primary care, they could go to a General Practitioner(GP).[3] They arrive at the GP with some extra family history information. Dee has a brother with Fragile X and she has not had premutation/carrier testing. She should have it, being at 50% risk. Likewise, the CF test for Aidan becomes important to see if their fetus is at high risk of CF. Therefore, they have the relevant genetic tests and come back a few days later to receive their results. She is given the all clear for Fragile X, but he turns out to be a CF carrier, so they are offered prenatal genetic testing in the form of CVS for the 1 in 4 risk of CF in the fetus. Population screening has led to genetic testing and implications for this nuclear family and perhaps their relatives.

Other information that came up at the return visit to the GP was the fact that Aidan was a smoker and could not give up, ignoring all the public health promotions about quitting, even though his father had died of coronary artery disease (CAD). The GP decides to take the opportunity to screen Aidan for susceptibility to CAD by checking his Apo E status.[4] He can do that right now by giving a cheek brush swab, unlike a lipid profile test which he would have to ask Aidan to fast for and come back another time for a blood test. Aidan hates going to the doctor, so, although Apo E is not considered any more beneficial than a lipid profile, if he does have the susceptibility Apo E genotype, perhaps that information will encourage him to quit smoking and choose a healthier diet. Many argue that this is not the way genetic screening should be used in a situation where there are perfectly good non-genetic interventions,[5] but maybe it will be the most opportune way sometimes. Of course, he could incidentally discover that his Apo E genotype puts him at increased risk of Alzheimer's disease.[6] What is he going to do about this at his young age?

So the apparently straightforward population screens this couple had (perhaps only a week ago) have raised many issues for this apparently healthy young couple.

Without a doubt, **screening can be beneficial**:

- firstly, to prevent or delay onset of an adverse health condition by identifying people at risk of developing a genetic disorder, thereby allowing for lifestyle, behavioural and or medical interventions.
- secondly, reproductive options become available
- and thirdly, screening may help to decrease social and financial burdens, both on these individuals and society.

3.   THE NINE CONCERNS

But to unbalance these benefits are many risks; perhaps "concerns" is a better label because the degree of importance and degree of resolution of these risks varies from country to country, jurisdiction to jurisdiction and in different models of healthcare. All have an impact on optimal performance of the screen or on the population as a result of screening—and they will not go away. They are the reasons why so few population screening programmes have been fully embraced by communities. Nine concerns will highlight the complexities related to implementation of population screening and the care with which it must be done. They are reviewed in more detail elsewhere.[7]

3.1   *Probabilistic Nature of Genetic Screening Information*

a) Screening is screening, not diagnosis, and there will invariably be false positives and false negatives until we have microarrays that cover all possible pathological mutations and variations, to represent the diversity of genotypes in different populations.[8] In the Victorian population, the screening panel for CF is for 12 mutations, with an average sensitivity of nearly 85%,[9] but to get closer to 100% sensitivity in our heterogeneous population, we would need to look at 100 mutations and have an array to put them on. This may not be too far away.

b) What is the predictive power of the information obtained, especially when the penetrance is unknown, nor the influence of other genes and environment? Screening provides a horoscope, not a map.

c) Rigorous evaluation guidelines emphasise the importance of analytical validity, clinical validity, utility, plus ethical and social implications to make sure these test characteristics, as well as the limitations of the results, are well understood.[10]

3.2   *Non-Modifiable Risk Factors*

At the individual level is the risk that people will not act upon the genetic information they receive because they perceive that nothing can be done about changing outcomes related to genes.[11] Quite clearly, no screening programme is going to be effective unless the majority of those screened fall into the "healthy behaviour" type: i.e. if a person discovers he or she is at increased risk, he or she must take extra care attending to health factors over which individuals have some control or, if at low risk, understand that it is still important to look after one's health in whatever way possible. "Unhealthy behaviour" in these two categories is to do nothing because of a sense of powerlessness or to take the attitude that one is in the clear and can do whatever one likes.

3.3   *Fear of Eugenics*

Genetic technologies have produced methods very much more powerful than the old eugenic methods of sterilisation and abortion to determine which genes are passed onto the next generation.[12] One of the most talked about concerns in the broader community is the fear that use of genetic information at the population level through routine screening (especially that related to prenatal screening), without consideration of basic human rights such as informed choice, is a form of eugenics. Prenatal and other screening is promoted as being done with informed choice, but choice may be constrained by service availability, incomplete information, or health professional attitudes. Additionally, choice may not be "free"[13] because, although screening is not mandatory (except some newborn screening in the US), subtle societal and peer pressures take over and some people feel they should have a test.

3.4   *Individuals Versus Populations*

There is a tension between managing genetic information relevant to an individual or within a family context and improving the health of the population.[14] Population screening strategies must encompass the individual views, particularly those related to individual privacy, which rise to the surface in the light of genetic information being obtained. This can mean that the public collective good and/or that of the family may come second.

3.5   *Security of DNA*

What can happen to a tiny drop of blood, swab of cheek, or follicle of hair, all of which are able to be used over and over again? How will DNA and the information obtained from it be stored and protected? These are big questions that experts have been addressing.

3.6   *Discrimination*

There has been much talk of the possibility of genetically susceptible population subgroups being identified, categorised, marginalised or discriminated against in various ways in non-clinical settings—the creation of the "genetic underclass." Family relationships, insurance (life, travel and health), employment, finance, adoption, migration, in the courtroom: are all examples of places where discrimination may occur.[15]

3.7   *Resource Allocation*

An emphasis on genetic liability to disease and disability may mean less funding for interventions addressing the social and other determinants of health. The allocation of limited, competitive public funding in the emerging world may be seen as more serious because priorities are more apparent for social and environmental problems there.

Second, there will not be enough money to fund the potential required investment into genetic service provision in a climate of very competitive funding for all healthcare services. There would undoubtedly be a need for a new or expanded service of genetic community workers.[16]

The third concern is that related to the potential cumulative effect of funding—the snowball effect. How big will the funding needs become and how will reimbursements be prioritised?

3.8   *Commercial Imperative*

There is the possibility of commercial returns for investment in genetic screening, particularly for pharmaceutical and biotechnology industries. Aggressive marketing may compete with any national approach to screening and cause inequities in access and affordability. Commercial advertising about what a screen can tell someone can undermine the doctor/patient relationship and lead to unreasonable demands on the doctor. Guidelines relevant to the laboratory and consumer relationship are being developed in this new ethical and regulatory domain.[17]

3.9   *Understanding and Education*

Last but not least, genetics needs a specialised language and, as many people have only studied basic classroom genetics, there are misconceptions about what genetics can do as well as an inability to evaluate the credibility of the genetic information. Responsible journalism will contribute to better understanding and provide perspective to the messages related to advances in genetics,[18] but education across community sectors is the key to progress. This is not just necessary for the lay population, but also for health professionals, the police, lawyers, government and even scientists!

4.   COMMUNITY PREPAREDNESS

Having described briefly the concerns for the community in regards to implementation of population genetic screening, the remainder of the chapter is about preparing the community for population screening. Our community can be divided up neatly into four sectors: consumers, health professionals, researchers and government.

4.1   *Consumer Preparedness.*

We live in a very complicated, risk-filled world, and we need to prepare the next generation now. Along with sex education, a basic health education curriculum should include information related to nutrition (e.g. folate), smoking, alcohol, drugs and an understanding of genetic risk and uncertainty, reproductive choices, and the importance of family history.

Those who are already candidates for population screening need different educational strategies and tools that are appropriate for different screening approaches. For example:

a)   Decision aids can provide information and help in decision making, especially if there is a choice of ways to be screened, as in antenatal screening
      or
b)   Face to face group (one to many) education programmes.

The following are examples from Australia of these two ways of tailoring methods of providing information for consumer preparedness.

We have recently completed a study involving the design and evaluation of a decision aid for prenatal testing of fetal abnormalities, known as the ADEPT study.[19] This three year cluster randomised controlled trial, funded by the Australian National Health and Medical Research Council, has recently finished and found that informed choice, as measured by Multiple Measure of Informed Choice[20] was improved by 70% in the decision aid arm of the trial, compared with a group who just received a pamphlet. However, only 42% of women in the decision aid group made an informed choice, compared to 32% in the pamphlet group. We believe decision aids have a lot of potential, but this study also raises the question as to whether informed choice will ever be possible and even of what exactly "informed choice" is.

In regards to group education, we developed a protocol, including a detailed education session, for people in the workplace to whom we offered hereditary haemochromatosis (HH) screening. Of those who attended the education session, 95% went on to be screened. The main study was published in *The Lancet* last year[21] and one on educational outcomes was published this year in *Clinical Genetics*.[22] The questionnaire responses of the 11 307 participants showed that 90% had good knowledge of the clinical concepts (aetiology, treatment options) and about 60% of genetic concepts (penetrance, genetic heterogeneity), and the 47 homozygotes we detected retained this information better than a control group.

The main problem with the workplace screening programme was that there was only 10% uptake overall. A questionnaire given to non-attendees found that there was lack of awareness of the programme, or not enough time to attend—these being indicators of disinterest rather than opposition. Interestingly, when the project officer went back to a workplace she had been to before, participation rose to 50%. It has been decided to try a different population, seen perhaps as a more captive population. To this end, this HH screening programme is being adapted for adolescents in schools,[23] and is basing its approach on the experience in Australia with Tay Sachs screening in schools.[24]

4.2   *Health Professional Preparedness*

What about heath professionals in primary care,[25] obstetrics and gynaecology, midwifery, paediatrics? Again, the crux of moving forward is education and professional development in areas such as communication of genetic information and principles of genetic risk, knowledge of the availability of screening tests and their characteristics, so that relevant choices can be offered. Recognition of the family history as a very important screening tool is essential, and there is a need for standardised, validated approaches for taking a family history.[26]

Of relevance to health professional preparedness were some outcomes of focus groups for the ADEPT study. When developing the decision aid, there were focus groups with GPs, and the themes that came out in regards to antenatal screening were time pressures, making the information real, being the gatekeeper of information, bearing bad news, facing moral imperatives, and dealing with problems inherent in "the system." Many of these issues apply to any population screening effort and must be addressed in preparation of a way forward. One not listed here because it is not so relevant to antenatal screening for Down syndrome is the impact on other family members and how to have them informed of their possible genetic risk. This is extremely relevant for other types of genetic conditions for which screening may be implemented.

In regards to helping health professionals find relevance for genetics within their practice, there is a major venture in Australia currently underway to produce a comprehensive and comprehensible resource funded by government and a biotechnology organisation. This is currently in draft form, going through a major public and professional consultation process. In addition, a much needed basic approach to education is to change undergraduate and graduate curricula to include more genetics. The new Human Genetics Advisory Committee in Australia has devised a "road map" for this and is in consultation with deans of a number of relevant university faculties. In the UK, there is the National Genetics Education and Development Centre and, in the US, the National Coalition of Health Professional Education in Genetics and many more global education initiatives.

4.3   Researcher Preparedness

There is quite generous funding for research in the area of complex disease genetic susceptibility and a growing amount of banked DNA for identification of important associations. It is imperative that association studies are repeated and that both positive and negative findings, and the related phenotypes, are reported.

What we need more of are advances in centralised databases to collect all the information coming out of these studies: e.g the Human Variome Project (HVP), which plans to develop standardised tools necessary for complete and systematic collection of variation, the phenotype and the studies performed.[27] This complements other collections such as the HapMap[28] and HuGENet[29] which all tackle collections from different angles: the HVP will collect Locus Specific Database mutation information; the HapMap stems from complete sequencing of common single nucleotide polymorphisms (SNPs) in a number of different individuals and attempts to show how these supposedly neutral variations are organised across the entire genome; and HuGENet coordinates and disseminates human genome epidemiologic information, such as genotype prevalence in different populations, magnitude of specific disease risk associations, validity and impact of genetic tests in different populations.

More funding support is needed for public health monitoring, evaluation and quality control. As already mentioned, there are excellent models for evaluation, as well as a new focus on health behaviour, the psychosocial impact, and cost benefit as per the 1998 revision of the WHO guidelines on genetic screening.[30]

Researchers must recognise when it is the right time to apply their research findings and with whom to engage to implement a screening strategy most effectively. The Evaluation of Genomic Applications in Practice and Prevention (EGAPP) model from the CDC is a systematic process for conducting evidence-based evaluations of genetic testing in transition[31] and will hopefully be adopted on a wide-scale.

4.4   *Government Preparedness*

If governments are to be prepared and understand the benefits and risks of screening they must:

a)   Develop a framework for policy decision-making that uses their peak expert bodies to advise them, and call on professional organisations and researchers, industry, academia, genetic service providers and special interest groups.

b) Engage those not historically involved in genetic service provision such as professionals in maternal and child health, disability services, chronic disease.

c) Follow up policy decisions with authoritative guidelines and legislation.

d) Address concepts, not just (dare I say it) vote catching issues of immediate relevance. A prime concept is avoidance of further health disparities, with genetic screening having enormous potential to make the gap wider.

Now more than ever before, it is important for governing bodies to engage all players and have open dialogue to find some common ground to clarify the goals of screening populations.

5.   BACK TO DEE & AIDAN CHIP, NOW IN THE YEAR 2020

So, what is the reality for Dee and Aidan in 2020? They will probably be able to pay $1000US to get a complete DNA profile in 2020[32]—a personalised genetic risk assessment for individuals who have given generic consent for all this information to be obtained. Someone is going to have to explain what it means for them and provide services to follow up and manage their future health. At the same time, the proposed pregnancy could be the result of a very careful screen of embryos or be tested during very early fetal development, using the RIP-OFF micro chip = Risk-Involved Predisposition of OFFspring chip to give the DNA profile. I think this chapter has demonstrated that there are many barriers to this being the reality.

6.   CONCLUSION

There are many possible factors and levels of risk to consider for individuals, the population, the health service and governance. The risks and benefits must be evaluated in real time to account for constant advances in identifying valid and useful associations and in determining the predictive power of genetic information. There is a need to continually evaluate, consider and reconsider different aspects of and models for implementation of population genetic screening.

Education of all stakeholders is imperative to build community understanding and trust. This will allow for free informed decision making, not just about the test process, but about what may follow and how to access appropriate health care. It is imperative to start as early as possible in schools, to evaluate different strategies, in different settings with different populations and to hold global meetings to share experiences, in readiness for 2020.

REFERENCES

1. B. Godard, et al., *Population Genetic Screening Programmes: Principles, Techniques, Practices, and Policies*, 11 EUR. J. HUM. GENET. (supplement) S49 (2003).

2. M. B. Delatycki et al., *Use of Community Genetic Screening to Prevent HFE-Associated Hereditary Haemochromatosis*, 366 LANCET 314 (2005).

3. N. Qureshi et al., *Timeline: Raising the Profile of Genetics in Primary Care*, 5 NAT. REV. GENET. 783 (2004).

4. C. Wang et al., *Combined Effects of ApoE-CI-CII Cluster and LDL-R Gene Polymorphisms on Chromosome 19 and Coronary Artery Disease Risk*, 209 INT. J. HYG. ENVIRON. HEALTH 265 (2006).

5. J. E. Eichner et al., *Apolipoprotein E Polymorphism and Cardiovascular Disease: A HuGE Review*, 155 AM. J. EPIDEMIOL. 487 (2002).

6. L. A. Farrer et al., *Effects of Age, Sex, and Ethnicity on the Association Between Apolipoprotein E Genotype and Alzheimer Disease. A Meta-Analysis. APOE and Alzheimer Disease Meta Analysis Consortium*, 278 J. A. M. A. 1349 (1997).

7. J. L. Halliday et al., *Genetics and Public Health – Evolution, or Revolution?*, 58 J. EPIDEMIOL. COMMUNITY HEALTH 894-9 (2004).

8. N. S. Green & K. A. Pass, *Neonatal Screening by DNA Microarray: Spots and Chips*, 6 NAT. REV. GENET. 147 (2005).

9. R. J. Massie et al., *Screening Couples for Cystic Fibrosis Carrier Status: Why Are we Waiting?*, 183 MED. J. AUST. 501 (2005).

10. CDC GaDP, ACCE: A CDC-Sponsored Project Carried out by the Foundation of Blood Research, http://www.cdc.gov/genomics/gtesting/ACCE.htm.

11. T. M. Marteau & J. Weinman, *Self-Regulation and the Behavioural Response to DNA Risk Information: A Theoretical Analysis and Framework for Future Research*, 62 SOC. SCI. MED. 1360 (2006).

12. D. J. Galton, *Eugenics: Some Lessons from the Past*, 10 REPROD. BIOMED. ONLINE (supplement) 133 (2005).

13. T. Shakespeare, *Choices and Rights: Eugenics, Genetics and Disability Equality*, 13 DISABILITY & SOCIETY 665 (1998).

14. J. C. Thomas et al., *Genomics and the Public Health Code of Ethics*, 95 AM. J. PUBLIC HEALTH 2139 (2005).

15. Australian Law Reform Commission, Essentially Yours – The protection of Human Genetic Information in Australia (2003).

16. Qureshi et al., *supra* note 3.

17. W. Burke & R. L. Zimmern, *Ensuring the Appropriate Use of Genetic Tests*, 5 NAT. REV. GENET. 955 (2004).

18. T. M. Bubela & T. A. Caulfield, *Do the Print Media "Hype" Genetic Research? A Comparison of Newspaper Stories and Peer-Reviewed Research Papers*, 170 C. M. A. J. 1399 (2004).

19. C. Nagle et al., *Evaluation of a Decision Aid for Prenatal Testing of Fetal Abnormalities: A Cluster Randomised Trial*, 9 B. M. C. PUBLIC HEALTH 96 (2006).

20. S. Michie et al., *The Multi-Dimensional Measure of Informed Choice: A Validation Study*, 48 PATIENT EDUC. COUNS. 871 (2002).

21. Delatycki et al., *supra* note 2.

22. A. E. Nisselle et al., *Educational Outcomes of a Workplace Screening Program for Genetic Susceptibility to Hemochromatosis*, 69 CLIN. GENET. 163 (2006).

23. A. A. Gason et al., *Genetic Susceptibility Screening in Schools: Attitudes of the School Community towards Hereditary Haemochromatosis*, 67 CLIN. GENET. 166 (2005).

24. A. A. Gason et al., *Tay Sachs Disease Carrier Screening in Schools: Educational Alternatives and Cheekbrush Sampling*, 7 GENET. MED. 626 (2005).

25. Qureshi et al., *supra* note 3.

26. M. J. Khoury et al., *Do we Need Genomic Research for the Prevention of Common Diseases with Environmental Causes?*, 161 AM. J. EPIDEMIOL. 799 (2005).

27. R. G. Cotton et al., *Locus-Specific Databases: From Ethical Principles to Practice*, 26 HUM. MUTAT. 489 (2005); R. G. Cotton & C. R. Scriver, *Human Mutation Databases*, 2 HUM. GENOMICS 272 (2006) (author reply).

28. HapMap, *A Haplotype Map of the Human Genome*, 437 NATURE 1299 (2005).

29. HuGENet , http://www.cdc.gov/genomics/hugenet/default.htm.

30. V. Goel, *Appraising Organised Screening Programmes for Testing for Genetic Susceptibility to Cancer*, 322 B. M. J. 1174 (2001).

31. CDC, Genomics and Disease Prevention, http://www.cdc.gov/genomics/activities/file/print/egapp.pdf (June 2006).

32. G. M. Church, *Genomes for All*, 294 SCI. AM. 46 (2006).

PART II

BALANCING INTERESTS
IN PUBLIC HEALTH GENOMICS

Introduction: Of Genomics and Public Health: Building Public "Goods"?[*]

Bartha Maria KNOPPERS

Faculté de droit, Centre de recherche en droit public, Université de Montréal

"Global public goods favor the mechanism of public information resources and free and open communication therein. Global public goods once produced should benefit all. Like a clean environment, knowledge about human health has no one institutional home. Like the gene pool at the level of the species being considered the common heritage of humanity, so genomic databases while recognizing the initial contribution of individual participants and of individual researchers or commercial investors should also account for the needs of present and future generations and foster and promote international collaboration."[1]

All signs point to the potential for the Human Genome Project to provide tools for the translation of genomic knowledge to clinical diagnosis, with implications for every level of the health care system.[2] Indeed, "[g]enomics is inspiring the development of very large longitudinal cohort studies and even studies of entire populations to establish repositories of biological materials ('biobanks') for discovery and characterization of genes associated with common

[*]. "Of genomics and public health : Building public "goods"?" – Reprinted from, CMAJ 08-Nov-05; 173(10), Page(s) 1185-1186 by permission of the publisher.

diseases."[3] With these "biobanks," an important advance in human genetics will be the identification and characterization of numerous common genetic variants at specific loci that increase or decrease the risks for various diseases singly and in combination with other genes and with various chemical, physical, infectious, pharmacologic and social factors. Yet, when applied to such population studies and to the ensuing accompanying genomic databases, current consent and privacy mechanisms may limit the use of these biobanks for public health research.[4]

Although the publicly available sequence map of the human genome was preceded by other international collaborative efforts[5] such as the mutation database initiative and, more recently, by the International Haplotype project, these essential scientific building blocks of understanding raise only limited privacy concerns.[6] More problematic are the privacy issues facing population banks that study genotype and phenotype interaction.[7] Currently still under construction, these human genetic research databases will constitute an immense public resource.[8]

Coupling human genomic databases with databases of pathogens yields the promise of a strengthened scientific basis for the primary and secondary prevention of disease. Combined with understanding of environmental factors, it will eventually provide the basis for programs of health promotion and disease prevention, when public health powers permit.

Norms for the emergence of a new paradigm for public health interventions must be informed by issues beyond the legal and ethical parameters of autonomy and privacy.[9] Indeed, the fundamental reason why contemporary medical ethics has so little to say about public health is that its focus on individual autonomy suggests that all compulsion for the sake of health is wrong. Yet "many public health measures must be compulsory if they are to be effective."[10] Thinking at the level of populations or groups requires a vetting of current ethical and legal principles and the development of a concept of the public good or of "common" goods.[11]

Privacy directives in Europe, laws in the United States and guidelines in Canada often treat personal genetic information as distinct from medical and personal data. Classical approaches to public health are based on the model of epidemic control, and the rise of autonomy and privacy legislation in the last decades has left little room for ongoing surveillance.[12] In short, genomic databases are pulled under this "genetic privacy" umbrella even when they are limited to the study of genomic variation (e.g., HapMap [www.hapmap.org], CARTaGENE [www.cartagene.qc.ca]). Such databases can range from descriptions of sequences, to annotated and curated databases, to disease-specific and, finally, longitudinal population databases such as the United Kingdom biobank (www.biobank.ac.uk). While basically oriented toward the building of scientific infrastructures and resources on genomic variation rather than individual disease-oriented studies on specific cohorts, there is no doubt that their potential usefulness for public health surveillance of genomic susceptibility to diseases is immense.

The concept of public goods has its roots in the 18th century. Hume coined the expression "providing for the 'common good'" in his Treatise on Human Nature (1739). Two main qualities exemplify "pure" public goods: its benefits are nonrivalrous in consumption (i.e., one person or group's use does not preclude another person or group's use of a public good) and nonexcludable (i.e., no one can be excluded from benefiting from a public good). Likewise, the "benefits of epidemiological intelligence are nondivisible for all countries."[13]

Ultimately, humanity as a whole should be the beneficiary of global public goods. The qualifying mark of a global public good is that it meets the needs of present generations without jeopardizing those of future generations.[14] It is the latter quality together with those of non-rivalry and non-excludability that led the HUGO (The Human Genome Organization) Ethics Committee in its 2002 Statement on Human Genomic Databases[15] to take the position on primary genomic sequences that:

1. Human genomic databases are global public goods. (a) Knowledge useful to human health belongs to humanity. (b)

> Human genomic databases are a public resource. (c) All humans should share in and have access to the benefits of databases.[16]

Policy development in this area must take contextual and cultural factors into consideration.[17] To avoid untoward effects, genetic research that identifies differential risks in populations requires special consideration before they are incorporated into laws, regulations or public health practices.[18] One of the underlying values of Canada's 2004 proposal for health protection renewal legislation is to "include public engagement in the decision-making process." Both collective and individual rights and interests are at stake in creating or accessing genomic databases for public health research.[19] It is also this "population focus [that] distinguishes public health from the clinical enterprise that is governed by the Hippocratic imperative with its focus on the individual patient."[20] It would be shortsighted indeed to fail to develop ethics for public health genomics, for the public funding of resources such as large genomic databases is ultimately premised on their usefulness in the public interest.

© 2005 Canadian Medical Association

REFERENCES

1. B. M. Knoppers & C. Fecteau, *Human Genomic Databases: A Global Public Good?*, 10 EUR. J. HEALTH LAW 27 (2003).
2. L. M. Beskow et al., *The Integration of Genomics into Public Health Research, Policy and Practice in the United States*, 4 COMMUNITY GENET. 2 (2001).
3. M. J. Khoury et al., *The Emergence of Epidemiology in the Genomics Age*, 33 INT. J. EPIDEMIOL. 936 (2004).
4. C. Verity & A. Nicoll, *Consent, Confidentiality and the Threat to Public Health Surveillance*, 324 B. M. J. 1210 (2002); Z. Lin et al., *Point of View: Approaches for Protecting Privacy in the Genomic Era*, 24 GENET. ENG. NEWS 8 (2004).
5. C. A. Semple, *Bases and Spaces: Resources on the Web for Accessing the Draft Human Genome – After Publication of the Draft*, 2 GENOME BIOL. 1 (2001); A. Marks & K. K. Steinberg, *The Ethics of Access to Online Genetic Databases: Private or Public?*, 2 AM. J. PHARMACOGENOMICS 207 (2002).
6. HapMap Consortium, *The International HapMap Project*, 426 NATURE 789 (2003); B. M. Knoppers & C. Laberge, *Ethical Guideposts for Allelic Variation Databases*, 15 HUM. MUTAT. 30 (2000).

7. A. Cambon-Thomsen, *The Social and Ethical Issues of Post-Genomic Human Biobanks,* 5 NAT. REV. GENET. 866 (2004).
8. Organisation for Economic Co-Operation and Development (OECD), OECD's Working Party on Biotechnology Held a Workshop on "Human Genetic Research Databases – Issues of Privacy and Security", Feb. 26-27, 2004, Tokyo, http://www.oecd.org/document/37/0,2340,en_2649_34537_31799845_1_1_1_1, 00.html.
9. R. Schabas, *Is Public Health Ethical?,* 93 CAN. J. PUBLIC HEALTH 98 (2002).
10. O. O'Neill, *Public Health or Clinical Ethics: Thinking beyond Borders,* 16 ETHICS INT. AFF. 35 (2004); S. S. Coughlin & T. L. Beauchamp, Ethics and Epidemiology (1996).
11. I. Kaul et al. eds., Global Public Goods (1999).
12. O'Neill, *supra* note 10.
13. Kaul et al., *supra* note 11.
14. *Id.*
15. HUGO Ethics Committee, *Statement on Human Genomic Databases* (2002), http://www.hugo-international.org/Statement_on_Human_Genomic_Databases. htm.
16. *Id.*
17. D. Choski & P. Kwiatkowski, *Ethical Challenges of Genomic Epidemiology in Developing Countries,* 1 GENOMICS SOC. POLICY. 1 (2005).
18. J. G. Hodge, *Ethical Issues Concerning Genetic Testing and Screening in Public Health,* 125 AM. J. MED. GENET. (part C) 66 (2004); P. A. Schulte, *Interpretation of Genetic Data for Medical and Public Health Uses, in* 277 Blood and Data: Ethical, Legal and Social Aspects of Human Genetic Databases (G. Arnason et al. eds., 2004).
19. M. Brazier & J. Harris, *Public Health and Private Lives,* 4 MED. LAW REV. 171 (1996).
20. Semple, *supra* note 5.

Privacy Issues in Public Health Genomics

Mark A. ROTHSTEIN

Institute for Bioethics, Health Policy and Law, University of Louisville School of Medicine

Herbert F. BOEHL

Institute for Bioethics, Health Policy and Law, University of Louisville School of Medicine

There are two main privacy issues surrounding public health genomics. The first issue is determining the proper scope of public health genomics. If the ambit of the field is too broad, then public health action is likely to extend into extremely sensitive matters, such as reproductive decision making, thereby encroaching upon an individuals' privacy interests. Thus, public health agencies could become involved in health care decisions best left to the individuals and their health care providers in the clinical setting. The second issue involves how best to protect informational privacy when individual genomic information is disclosed for public health purposes. For example, the aggregation of data in anonymous or deidentified form might better protect privacy, but the information is likely to be less valuable for public health surveillance or research.

Deciding on rules for confidentiality and disclosure of genomic information for public health purposes requires a difficult balancing of individual and communal interests. Further complicating the analysis is the rapid and ongoing development of new genomic technologies. This chapter reviews the contours of public health genomics and recommends a cautious approach to the use of genomics in public health.

1.   PRIVACY AND THE SCOPE OF PUBLIC HEALTH GENOMICS

There is some confusion regarding the definitions of key terms in this field. If, as this chapter argues, public health genomics should be narrowly circumscribed, it is essential to have sound definitions of the operative terms.

1.1   *Definitions*

Genomics is the use of genome-wide analytical tools to study the effect of genes, proteins, and other gene products on the biological processes of an organism. It differs from genetics in the scientific scope of the analyses and applications.

Privacy is the quality or state of being apart from company or observation, and it refers to one of the following categories of concern: (1) informational privacy concerns access to personal information; (2) physical privacy concerns access to persons and personal spaces; (3) decisional privacy concerns governmental or other third-party interference with personal choices; and (4) proprietary privacy concerns the appropriation and ownership of interests in human personality.[1] Public health genomics could implicate all four of the concerns of privacy, but it especially raises issues related to decisional privacy—i.e., government interference with personal health choices (e.g., mandatory newborn screening)— and informational privacy—access to and uses and disclosures of personal health information (e.g., public health surveillance).

Public health is a term for which there are numerous definitions. I have previously advocated a narrow definition of public health, as involving "public officials, acting pursuant to specific legal authority, and after balancing private rights and public interests, taking appropriate measures to protect the health of the public."[2] This definition may be considered a "governmental" definition of public health. By contrast, a broader definition is that "public health is what we, as a society, do collectively to assure the conditions for people to be healthy."[3] In my view, this "population health" definition of public health fails to distinguish between the roles of governmental and private actors, does not differentiate between measures to advance the health of individuals and the public, and fails to justify the possible use of coercive measures.[4] An even broader definition is that public health addresses all of the societal factors that affect health, including war, violence, poverty, economic development, income distribution, natural resources, diet and lifestyle, health-care infrastructure, overpopulation, and civil rights.[5] In my view, this "human rights" definition of public health is imprecise and overbroad, focuses on areas beyond the expertise of public health practitioners, makes public health too politicized, and diverts attention from traditional public health issues.[6]

At least in the United States, public health is a legal term of art, and it refers to specifically delineated governmental powers, rights, duties, and responsibilities. Public health authority is based on express constitutional, statutory, and regulatory provisions.[7] Public health applies to specific institutions and individuals, such as public health departments and public health officials. Governments at all levels have a role to play in individual health, population health, and human rights, but it is a different role in kind and degree from its role in public health.

Adoption of the "governmental" definition of public health has two implications for privacy and confidentiality. First, it suggests that a heightened standard of societal need should be required before compulsory data sharing may be ordered by the government. Second, limiting disclosure of genetic information to official uses may have the effect of preventing excessive disclosures to other third parties whose activities fall outside of the definition of public health.

1.2   Public Health Genomics and Clinical Interventions

Another area of confusion is the relationship between public health and individual clinical interventions, a misunderstanding caused, at least in part, by the mistaken belief that government programmes that provide health care to indigent populations is "public health." To eliminate this confusion, it is helpful to note the three criteria that distinguish public health from individual clinical care, including clinical care provided by a public entity. First, public health acts when the health of the population is threatened. Although the prototypical public health activity is infectious disease control, the threat to the public need not be based on horizontal, person-to-person transmission, such as with environmental health hazards. Second, public health relies on the unique powers and expertise of the government. For example, disease reporting and surveillance are responsibilities of government acting through the public health system. Third, public health action by the government is more efficient or more likely to produce an effective intervention. Newborn screening would be an example where public health action is justified in providing the framework for supporting an important aspect of individual health care.

Based on these definitions and considerations, public health genomics is the use of genome wide analytical tools (or data derived from the application of those tools) by public officials, who are acting pursuant to specific legal authority to protect the health of the public.

It is important to note that the integration of genomics (and genetics) into public health interventions must be done carefully and with discretion because the values underlying public health are quite different from those underlying genomics and genetics. The exercise of public health authority not only involves governmental action, but it also means the possible use of coercive powers to enforce governmental objectives. Public health action is based on utilitarian and communitarian ethics, under which societal interests take precedence over individual interests. An example is the imposition of quarantine to fight the spread of an epidemic. By contrast, the dominant social values of genomics and genetics are autonomy, privacy, and reproductive freedom. Politically, it is more libertarian than communitarian. An example is the traditional client-centered, nondirective approach that has become a hallmark of genetic counselling.[8] The role of the genetic counsellor or medical geneticist is to educate the individual and to provide options, but the ultimate decision rests with the individual.

1.3   *Values*

The inherent conflict between the values underlying public health and those underlying genetics and genomics strongly suggests that any undertaking in the field of public health genomics should be approached with great care. This need for caution is underscored by the history of eugenics, which represented a failed and discredited attempt at reproductive, public health genetics. Today, members of the public overwhelmingly are concerned about the applications to individuals of genetic technologies (especially with regard to reproduction) to achieve social goals. Eugenics continues to be a dark cloud over public health genomics, and we ignore this fact at our peril.

1.4   *Hemochromatosis*

How do these values, interests, and concerns play out in practice? A good example is screening for hereditary hemochromatosis. Hemochromatosis is an autosomal recessive disorder characterized by excess iron absorption and deposition in tissue.[9] If left untreated, it can result in liver disease, diabetes, cardiomyopathy, and other serious disorders. Before the advent of genetic testing, a definitive diagnosis of hemochromatosis required a liver biopsy or other invasive testing, and there was no way to identify individuals who were presymptomatic. Furthermore, the recessive inheritance pattern of the disorder often did not make at-risk individuals or their physicians aware of their condition before the internal organ damage caused by excessive iron accumulation.

At first glance, hemochromatosis seems like an ideal candidate for public health intervention. The condition is relatively common, with as many as one million affected persons in the United States;[10] prompt detection can prevent harm; there is a cheap, easy, and effective therapy (periodic phlebotomy); there is a cheap, simple test; and the most affected group is not medically or socially vulnerable (hemochromatosis disproportionately affects white males).[11] Nevertheless, most experts and consensus panels that consider the issue have concluded that it is premature to offer population screening because of inadequate data on prevalence and penetrance of the most common mutations, lack of laboratory standardization, lack of agreement on optimal care for asymptomatic mutation carriers, and the fear that individuals testing positive will be subject to stigmatization and discrimination.[12]

In the extensive literature on hemochromatosis, there has been some inconsistency about whether efforts to promote population screening should be considered "public health" or "population health." This determination will affect the role of government agencies *vis à vis* individual clinicians, medical societies, medical specialty groups, and payers in implementing testing. Applying the three factors mentioned above (threat to population health, unique governmental powers and expertise, and government intervention as more efficient and effective), it is clear that population screening for hemochromatosis is not a proper public health activity. It is an open question whether, in the future, a compelling case can be made for routine hemochromatosis testing in clinical settings. If such data could be marshalled, however, then hemochromatosis testing should be implemented in a manner more akin to cholesterol and PSA testing (i.e., under individual or population health principles) than to newborn screening or immunizations (i.e., proper domains of a government-directed public health programme).

1.5   *Factor V Leiden*

Another example of the type of analysis required before implementation of a public health (or population health) genetics programme is Factor V Leiden. Factor V Leiden is a genetic mutation that results in thrombophilia or an increased propensity for the formation of blood clots.[13] A synergistic relationship between Factor V Leiden and the use of oral contraceptives has been observed. Whereas the use of oral contraceptives alone increases the risk of venous thrombosis by a factor of about 4, and the presence of Factor V Leiden alone increases the risk by a factor of about 7, their joint effect is to increase risk by a factor of more than 30.[14] This substantially increased relative risk raised the question of whether screening for Factor V Leiden should be routine before prescribing oral contraceptives.

Despite the high relative risk, there is still a low absolute risk of venous thrombosis (about 28 per 10 000 person-years) among women with Factor V Leiden who take oral contraceptives, and there is a low mortality rate among young women. It has been estimated that more than half a million women would need to be screened for Factor V Leiden, resulting in tens of thousands of women being denied oral contraceptives, to prevent a single death. "In addition to medical and financial considerations, there are issues related to the quality of life, the risk of illness and death from unwanted pregnancy, and concern about possible discrimination by insurance companies."[15]

Factor V Leiden indicates the complexities of decision making about the introduction of routine genetic testing in clinical practice. Because the evidence does not support the introduction of Factor V Leiden screening in the clinical setting, the issue of screening under the aegis of public health need not be reached. Nevertheless, it is clear that Factor V Leiden provides another set of issues to consider, but still presents a much less compelling case than hemochromatosis for population-wide screening.

2.   INFORMATIONAL PRIVACY

2.1   *HIPAA*

In the United States, laws protecting the privacy of health information are extremely deferential to public health authorities. Thus, health care providers are permitted to disclose individually-identifiable health information to public health agencies for public health purposes without the consent or authorization of the individual. The primary federal law regulating health information is the Health Insurance Portability and Accountability Act (HIPAA).[16] Although the law was enacted to make group health benefits portable among employers, a provision of the law deals with the privacy of health information. The statute directed the Secretary of Health and Human Services (HHS) to issue regulations protecting the privacy of health information, and HHS has done so in its Standards for Privacy of Individually Identifiable Health Information (Privacy Rule).[17]

Under the Privacy Rule, a covered entity (including a health care provider) may not disclose individually-identifiable health information for purposes other than treatment, payment, or health care operations without the written authorization of the individual. There are several exceptions to this general rule and public health is the first one listed in the Privacy Rule. Under the public health provision of the Privacy Rule, a covered entity may disclose the following five types of protected health information to public health authorities without a written authorization from the individual: (1) information related to the reporting of disease, injury, vital events such as birth or death, and the conduct of public health surveillance, public health investigations, and public health interventions; (2) reports of child abuse or neglect (a separate provision deals with reports of domestic violence); (3) reports about the quality, safety, or effectiveness of products regulated by the Food and Drug Administration; (4) reports that a person may have been exposed to a communicable disease or may otherwise be at risk of contracting or spreading a disease or condition; and (5) reports about an employee who may have contracted a work-related illness.[18]

This is another area where the definition of public health comes into play. The Privacy Rule permits disclosure of health information only to public health authorities for a public health purpose. Actions that could be considered coming under a narrow definition of "public health genomics" probably would also come under "public health interventions" in the Privacy Rule. Broader definitions of "public health genomics," however, could include issues beyond the scope of the Privacy Rule provision for public health reporting. Furthermore, the Privacy Rule *permits* such disclosures without an authorization; it does not require an entity to make any disclosures. Because health care providers are only required to make disclosures mandated by law, they should be scrupulous in protecting the confidentiality of genomic information, regardless of any asserted public health justifications, in the absence of a legal requirement for disclosure.

2.2 *Biosurveillance*

The greatest system-wide challenge to health privacy is the development of networks of interoperable, longitudinal, comprehensive electronic health records. In the United States, the Nationwide Health Information Network (HNIN) is being developed under the leadership of HHS.[19] Although the specific structure of the NHIN has yet to be determined, the avowed purposes of the NHIN are to improve safety, efficacy, efficiency, and quality of health care. Similar networks are in various stages of development in Australia, Canada, Denmark, the United Kingdom, and other countries.[20]

Besides improvement in the delivery of health care, the NHIN has the potential to improve public health epidemiology, including genetic epidemiology. The NHIN should make it much easier to correlate genotype and phenotype, to obtain prevalence data on rare disorders, and to compile data on treatment outcomes for genetic disorders. To the extent that these research activities involve individually-identifiable health information, then both the Privacy Rule and the Federal Policy for the Protection of Human Subjects (the Common Rule)[21] would need to be satisfied with regard to informed consent and other human subjects protections.

The NHIN also is being promoted as a means for engaging in real-time biosurveillance involving natural (e.g., influenza) and man-made (e.g., bioterrorism) health threats. Genomic biosurveillance could be part of the mix, as information related to variability in sensitivity, degree of response, and effectiveness of therapies also would be of interest to public health and homeland security officials. Thus, genomics, informatics, and biosurveillance could combine to present challenges to privacy.

In the rush to integrate these technologies, I believe that some public officials have failed to exhibit the appropriate level of caution. Just because a technological "advance" is possible does not mean that it ought to be implemented. Similarly, although national security is obviously an important interest, its invocation should not be considered an abracadabra that justifies any undertaking, no matter how intrusive, without adequate scrutiny.

In my view, the following are essential conditions precedent before the nation adopts a system of electronic health records biosurveillance, including the use of genetic or genomic factors.

First, public officials need to make a *compelling* case of the need for such a system and that it would represent a substantial improvement over existing methods of public health surveillance and reporting. Such a determination should rely on the results of pilot projects and smaller scale start-up measures before engaging in a massive undertaking.

Second, the biosurveillance system should be the least intrusive possible, consistent with programme objectives. Disclosure policies should involve the minimum amount of data in the least identifiable form. Disclosing more information is not always necessary nor is it perceived as necessary by the public.

Third, there should be a mechanism for meaningful input into system design and implementation by all stakeholders, including state and local public health officials, health care providers, and members of the public. The input should be obtained before the system is put into place. Unfortunately, there has been exceedingly little public notice of, let alone involvement in, the proposed biosurveillance system.

Fourth, there should be public and professional education about the programme. Simply providing another inscrutable notice to ill and weary patients at a vulnerable time is inadequate. Health care professionals also need to understand their role in the system, and they need to be able to address the concerns of their patients.

Fifth, there should be an ongoing programme of oversight, assessment, and research to ensure that the system is meeting its objectives in the least intrusive manner. The research should be of a probing and critical nature, and it should be undertaken by disinterested parties. Aspects of the system that are problematic should be changed or discontinued.

2.3   Stigma and Discrimination

Before substantial amounts of effort and expense are devoted to protecting privacy and confidentiality in public health genomics, it is fair to consider the specific privacy and confidentiality interests at stake. Numerous public opinion surveys indicate that many members of the public are deeply concerned about genetic privacy,[22] although the nature of that concern is not always clear. Individuals are concerned about both the intangible and tangible aspects of a loss of genetic privacy. As to the former, individuals often suffer psychologically from having genetic information disclosed to family members, loved ones, colleagues, and friends, especially when the information involves certain sensitive or untreatable medical conditions. Disclosure of the information to strangers often carries even greater concerns about embarrassment, humiliation, and emotional distress.

The word stigma comes from the Greek and literally means the scar or mark left by a burning or cutting of the flesh. Today, stigma refers to a mark of shame or discredit. With genetic information, stigma not only attaches to personal genetic information, but from genetic information about relatives or even unrelated members of the same ethnic group. Thus, information about the genetic etiology of, for example, a parent's or sibling's mental illness might be considered stigmatic because it would have implications for the risk (future or reproductive) of currently unaffected relatives. Stigma also may attach in a more general way where, for example, researchers disclose that members of a certain ethnic group are at an increased risk of drug abuse, mental illness, or some other disorder. Unfortunately, history is filled with numerous examples of group-based stigma arising in the context of illness and disease.

Many individuals also are concerned about the potential for genetic discrimination that might occur if their genetic information were disclosed. There are essentially two concerns. First, many individuals are afraid that employers, insurers, and other third parties will misuse genetic information to make *inaccurate* predictions about an individual's health risks and thereby deny access to employment, insurance, or some other commercial relationship. Second, many individuals also are concerned that employers, insurers, and other third parties will correctly use genetic information to make *accurate* predictions about individuals' health risks, with the result that they are denied access to employment, insurance, or some other commercial relationship that they consider essential and to which they believe they should have reasonable access.[23]

Along with numerous other commentators, I have written at length about various aspects of genetic discrimination, and this is not the place to revisit those issues and arguments. For the purposes of public health genomics, it is sufficient to say that a broad range of apprehensions subsumed under "stigma" and "discrimination" are sure to be raised by public health genomics if there is more widespread disclosure of genetic information.

3. CONCLUSION

Because the purpose of public health is to advance the well-being of the population, sometimes at the expense of individual interests, public health may conflict with the ethical tradition of autonomy that has become ingrained in genetics. Only by carefully limiting the scope of public health activities in genetics can the public be assured that the government is not unreasonably encroaching into some of the most sensitive and private realms of individual health.

New genomic and bioinformatics technologies also will substantially increase the amount of genetic and genomic information contained in health records. Stringently enforced confidentiality and security safeguards must apply to the protection of this information. At the same time, public officials should be required to present a compelling justification before individual genetic or genomic information is mandated to be disclosed for public health purposes, including for use in biosurveillance. Even then, such disclosures should be the minimum necessary to achieve the purpose of the disclosure and in the least identifiable form. As with other types of sensitive health information, vigilance in protecting the privacy of genetic and genomic information is especially warranted where, as is the case with public health genomics, the ostensible purpose for the disclosure is benign.

REFERENCES

1. A. L. Allen, *Genetic Privacy: Emerging Concepts and Values, in* Genetic Secrets: Protecting Privacy and Confidentiality in the Genetic Era 33 (M. A. Rothstein, ed. 1997).
2. M. A. Rothstein, *Rethinking the Meaning of Public Health,* 30 J.L. MED. & ETHICS 144 (2002).
3. Committee for the Study of the Future of Public Health, Institute of Medicine, The Future of Public Health 19 (1998).
4. Rothstein, *supra* note 2, at 145-46.
5. *Id.* at 144 & n.1.
6. *Id.* at 145.
7. *See generally* L. O. Gostin, Public Health Law: Power, Duty, Restraint (2000).

8. *See* A. L. Caplan et al., *Neutrality Is Not Morality: The Ethics of Genetic Counseling, in* Prescribing our Future: Ethical Consideration in Genetic Counseling (D. M. Bartels et al., eds., 1993).
9. D. L. Witte et al., *Hereditary Hemochromatosis,* 245 CLIN. CHIMICA ACTA 139 (1996).
10. *Id.*
11. M. E. Cogswell et al., *Iron Overload, Public Health, and Genetics: Evaluating Evidence for Hemochromatosis Screening*, 129 ANNALS INTERNAL MED. 971 (1998).
12. W. Burke et al., *Hereditary Hemochromatosis: Gene Discovery and Its Implications for Population-Based Screening,* 280 J. A. M. A. 172 (1998).
13. M. J. Khoury et al., *Population Screening in the Age of Genomic Medicine*, 348 NEW ENG. J. MED. 50 (2003).
14. *Id.* at 55.
15. *Id.*
16. 42 U.S.C. §§ 300gg-300gg-2 (2000).
17. 45 C.F.R. Parts 160, 164 (2004).
18. 45 C.F.R. § 164.512(b) (2004).
19. *See* National Committee on Vital and Health Statistics, Letter to Health and Human Services Secretary Mike Leavitt, June 26, 2006, http://www.ncvhs.hhs.gov/060622lt.htm.
20. *Id.*
21. 45 C.F.R. Part 46 (2004).
22. *See, e.g.,* M. A. Rothstein & C. A. Hornung, *Public Attitudes, in* Genetics and Life Insurance: Medical Underwriting and Social Policy (M. A. Rothstein ed., 2004).
23. *See* M. A. Rothstein & M. R. Anderlik, *What Is Genetic Discrimination and When and How Can It Be Prevented?,* 3 GENETICS IN MED. 354 (2001).

Balancing Private and Public Interests in Policy

Darren SHICKLE

Academic Unit of Public Health, Institute of Health Sciences and Public Health Research, University of Leeds, UK

Public policy frequently has to reconcile tensions between public and private interest, at times being paternalistic, while recognising the importance of privacy and autonomy and balancing the interests of some against those of others. Governments can also at times appear to be protecting the interests of private sector organisations or of industry over and above the interests of individual citizens.[1] However, individuals may benefit from government policies aimed at safeguarding the interests of industry if these organisations contribute to the strength of the economy, are a source of employment, or provide products/services that the consumer wants. This chapter explores some of these tensions in the context of public policy in relation to genetic testing and insurance.

1.   PUBLIC CONCERNS ABOUT GENETIC TESTING AND INSURANCE

The general public is generally concerned about the use of genetic information by the insurance industry. The situation with regards to insurance is somewhat different in the United States compared with most of Europe. Citizens of the United States are particularly concerned about issues relating to insurance and genetic testing because of the dependence of a majority of the population on private insurance for healthcare (either directly or via their employer). In

Europe, in comparison, healthcare is usually funded out of taxation or social insurance, but concerns still exist in relation to other forms of insurance, such as life insurance.

A Time magazine/CNN poll published in 2000 found that only about 20% of people said that genetic information should be available to insurance companies.[2]

In a survey conducted for the UK Human Genetics Commission, 62% of respondents thought that genetic information *could* be used for setting the level of insurance premiums, but only 8% of people thought that it *should* be used for this purpose.[3] 78% of respondents disagreed with a statement that "insurance companies should be able to ask to see the results of genetic tests to assess whether premiums should go up or down." When asked if it was "appropriate or inappropriate for an insurance company to know the results from a genetic test that an individual has already undertaken (for example, risk of Huntington's disease or a rare cancer) when considering an application": 35% thought it was appropriate for an application for health insurance; 33% for long-term care insurance; 30% for life insurance; 21% for motor insurance; 19% for pensions; 18% for travel insurance; and 6% for home contents insurance.

The loss of health insurance was the greatest concern about genetic testing amongst a cohort of families with Hereditary Nonpolyposis Colorectal Cancer.[4] Of 78 eligible women who declined BRCA1/2 testing in Michigan, 48 cited concerns about cost and insurance discrimination.[5] Geer et al.[6] interviewed 37 people who had declined genetic counselling for cancer. Impact on insurability of self and/or family members was the most frequent reason (41%) given for declining counselling. Matloff et al.[7] asked cancer genetic specialists what they would do if they were at a 50% risk of carrying a gene for hereditary breast/ovarian cancer or colon cancer. The majority (68%) said that they would not bill their insurance company for the genetic test and 26% would use an alias, because of fear of discrimination. A number of reports and surveys have also described examples of genetic discrimination by the insurance industry both in the United States[8] and the United Kingdom.[9]

2. THE BASIS OF INSURANCE

In the narrow sense, insurance is a contract between an individual and an insurance company. Individuals perceive that they are at risk of an event and wish to offset the negative consequences by some financial recompense. For example, drivers may purchase car insurance in case they are involved in an accident and are concerned that they could not afford to pay the repair bill. Thus, they are willing to pay a relatively small premium each year that they can afford, just in case they are faced with a repair bill that they cannot afford. Similarly, someone may seek health insurance in case they become ill and are no longer able to work and/or have medical/social care bills that have to be paid.

In 17th century England, a merchant with a ship to insure in case of loss at sea would request a 'broker' to take the policy from one wealthy merchant to another until the risk was fully covered. The broker's skill lay chiefly in ensuring that policies were underwritten only by people of sufficient financial integrity as any claim would be made against their personal fortune. In 1688, Edward Lloyd opened a coffee house in London, encouraging a clientele of ships' captains, merchants and ship owners. The coffee house earned a reputation for trustworthy shipping news and became recognised as the place for obtaining marine insurance. In 1769, some of Lloyd's more reputable customers broke away to set up a rival establishment. This was one of the first demonstrations of any community of interest among insurance underwriters with the subsequent development of a constitution and trust deed, and led to the organisation known as Lloyd's of London.[10]

Lloyd's of London is still characterised by syndicates of members who underwrite insurance policies from their personal wealth. More typically, an insurance contract is with a company owned by shareholders. This contract between an individual and a third party who is willing to share risk in exchange for a fee does not usually exist outside the context of many similar contractual arrangements. In order for an insurance company to cover any payments to clients who may have a legitimate claim, it must obtain premiums for other clients who subsequently do not need to make a claim.

3.   FORMS OF SOLIDARITY

Jørgen Husted[11] made a distinction between two basic meanings of solidarity: *communal solidarity* where a group of people have *a common interest* and *constitutive solidarity* where people *have an interest in common*.

He further subdivided communal solidarity into *group solidarity* and *moral solidarity*.

Within group solidarity, the common interest is the *cement* or *organising principle* of the group. The members have a common interest in the sense that what is good or harmful to this interest is (or, at least, is perceived to be) good or harmful to the individual, too. Husted gave various types of groups to which an individual could belong. For example, an ethnic minority, a profession, a creed, a unit of organised labour, or a local community. Group members demonstrate solidarity by standing by weak and needy members in the sense of *looking after one's own*. However, this pattern of behaviour is more than just helping people in need, which could be valuable in its own right as a form of moral responsibility, as it is implicit or even explicit that this form of solidarity is in the common interest. As Husted points out:

> "By recognizing its collective responsibility towards its needy members the group secures the loyalty of all members to the common cause and thus, also in this way, promotes it. In the same way the group is able to make legitimate demands on the individual to contribute their share to the lifting of the burden of the collective responsibility."

In situations where solidarity was not practiced for the common interest of an identifiable group, or only to a limited extent, Husted suggested that people may still demonstrate solidarity for the sake of the needy benefiting from it. In such circumstances, the act is out of individual moral responsibility rather than collective responsibility to a group. Instead of a defined group with shared aims and objectives, there is a more general bond between individuals, a sense of *sharing a common lot* and *recognising oneself in the other*. Husted suggested that the basic principle underpinning this form of solidarity is *making the other person's cause one's own* out of a sense of duty.

Husted identified the following as important forms of moral solidarity:

- Brotherhood (sisterhood) solidarity: For example, supporting others elsewhere in the world in disadvantaged political settings.
- Charitable solidarity (neighbourly love or philanthropic solidarity): Provision of help out of a feeling of *doing unto to others as they would want done unto them* if they were also in need.
- Social solidarity: Willingness of well-off citizens to help the poor and needy via income redistribution.
- Egalitarian solidarity: Provision of social goods, for example health care, according to need rather than ability to pay.
- Humanist solidarity: For example, humanitarian aid following a natural disaster or in a war zone or protests against oppression of others.

Unlike group solidarity where the focus is on what is in the best interests of the group, and moral responsibility where the focus is on the best interests of needy individuals, *constitutive solidarity* (or *alliance solidarity*) is focused on the interests of the individuals themselves. In this latter situation, individuals realise that the best way to advance their own individual interest is to form an alliance with others to establish some form of collective agreement specifying the expected contribution to the collective and what they can expect back in return. Husted gave two examples of this. Firstly, workers may come together within a trade union to strengthen their ability to negotiate with employers by increasing the threat of industrial action. Individual workers would have a weak negotiating position as the employer could 'pick them off' one-by-one. But if the entire workforce *stands together* and withholds their labour simultaneously, the impact would be greater. In return for paying their union fees and abiding by agreed industrial action, if necessary, the employee is likely to get better pay and working conditions. The other example of constitutive solidarity Husted described is *entrepreneurial solidarity*. In this form of solidarity, individuals come together as stockholders to establish a company, with the goal of the individual increasing his or her own capital. Similarly, farmers could form a cooperative, for example to share equipment, marketing or negotiation processes with suppliers and customers.

While Husted draws a distinction between *common interest* and *interest in common*, it may be more productive to categorise his three main forms of solidarity as *group, moral* and *constitutive*, according to the main interest being considered within each.

> Within group solidarity, the main focus is on the best interests of the group. The individual is part of the group and benefits if the group flourishes, but it is the collective interest that is the main concern.
> Within moral solidarity, the main focus is on third party individuals and doing things for them because it is the right thing to do. While there may be some expectation that others would act in the same way if the positions were reversed, in the pure sense of moral solidarity, the action is purely altruistic, and there is no expectation of personal reward of acting morally in doing the *right thing*.
> Within constitutive solidarity, the main focus is the individual themselves. They are working with other people, and so indirectly assisting others to advance their goals, but the focus is benefit to self.

4.   ADVERSE SELECTION

Insurance could be seen as a form of constitutive solidarity. Individuals seek insurance because they perceive that it is in their own interest to do so. However, they depend on other people also purchasing insurance in order to make a market in which it is worthwhile for insurance companies to operate. Thus, insurance company clients are indirectly working with other people (who they usually do not know) with whom they have an interest in common, i.e. concerns about the financial consequences of a particular risk.

One of the principles of entrepreneurial solidarity is that "individuals receive benefits (or have to accept losses) in proportion to their individual contribution."[12] Within the context of insurance, the benefit is offset risk, and the greater the risk that is offset, the greater is the contribution required, i.e. a higher premium must be paid. Actuarial data is crucial to the success of the insurance industry. Insurance companies assess the risk of any eventuality and the potential (financial) consequences. Based on past experience, the insurance company calculates the premium that an applicant needs to pay to provide 'cover' against injury or loss. If/when the insured event happens, the company pays out the agreed level of claim. In order for an insurance company to be profitable, overall, the total premiums paid by all of its customers must exceed the total claims which may need to be paid out.

The insurance industry is based on probabilities. Information about these probabilities is of interest to both the person seeking insurance and the insurer. If individuals know that they are definitely not at risk, or at very low risk, then they will not apply for insurance. If the insurance company knows that the insured event is definitely going to take place, then they will not want to accept the application. The exception to this is life insurance, where death will definitely take place, the probability assessment relates to whether life expectancy is shorter or longer than average, and, as in other forms of insurance, actuarial data can be used to inform this risk assessment.

Insurance companies become concerned in situations where applicants have privileged access to information that modifies the assessment of risk. This situation is called adverse selection.

The United States Actuarial Standards Board defines adverse selection as: "Actions taken by one party using risk characteristics or other information known to or suspected by that party that cause a financial disadvantage to the financial or personal security system *(sometimes referred to as antiselection)."*[13] The Actual Standards Board goes on to explain why this is potentially problematic:

> "Adverse selection may result from the design of the classification system, or may be the result of externally mandated constraints on risk classification. Classes that are overly broad may produce unexpected changes in the distribution of risk characteristics. For example, if an insurer chooses not to screen for a specific risk characteristic, or a jurisdiction precludes screening for that characteristic, this may result in individuals with the characteristic applying for coverage in greater numbers and/or amounts, leading to increased overall costs."[14]

Adverse selection is not a concept restricted to genetics, but it is in this context that the term is most commonly used by people from outside the insurance industry. Pokorski described the concern of the insurance industry arising from the use of genetic tests as a "worry about the potential rise in applications by people who are aware of information that affects their likelihood of making an early claim but who choose not to inform their insurance company."[15] Thus, an individual may undergo a genetic test, and if it shows that that they will/may develop that disease, they may seek insurance (or large value of coverage) without disclosing the test result, knowing that they are likely to want to make a claim on the policy sooner than the insurance company would expect based on the other actuarial data available to them.

Zick et al.[16] followed 148 cognitively normal people participating in a randomised clinical trial of genetic testing for Alzheimer's disease for one year after risk assessment and Apolipoprotein E genotype disclosure. People who tested positive were 5.76 times more likely to have altered their long-term care insurance than those who were not told that they had the Apo ε4 genotype, although there were no differences in health, life or disability insurance purchases.

Of course, if a predictive genetic test is negative, it would be in an applicant's best interests to disclose the test result, especially if they have a strong family history for that disease, in the hope that they would be offered a policy with a lower premium as their risk of making a claim would be reduced.

Selective disclosure of negative rather than positive predictive test results may seem like a 'victimless crime.' What is the harm in 'playing the rules of the game' to your advantage? Insurance companies may be perceived as corporate organisations who make large profits. One individual making an 'unexpected' claim may be perceived to have minimal impact on shareholders' dividends. However, it is not unreasonable for insurance companies and their shareholders to make a fair return on their investment. Thus, while one person using undisclosed material information is unlikely to have any significant impact, the cumulative effect of a number of such claims is likely to mean a rise in premium levels for other policy-holders.

The United Kingdom Human Genetics Advisory Commission Subgroup on Insurance described how:

> "adverse selection can occur when the distribution of risk in a pool of insured people is skewed adversely, e.g. when more high risk people find it worthwhile to take out insurance. This drives up the price of premiums, so that low risk people may be deterred from taking out policies and may withdraw – this leads to a vicious circle of worsening of the risk pool and increasing costs."[17]

The U.S. Actuarial Standards Board was concerned that "adverse selection can potentially threaten the long-term viability of a financial or personal security system."[18] While these concerns may be extreme, adverse selection in which one or a group of individuals seek personal advantage at the cost of higher premiums for others could be considered to be an abuse of solidarity.

5.   THE RESPONSE OF THE INSURANCE INDUSTRY TO GENETIC TESTING

Van Hoyweghen et al.[19] examined the debate within the insurance industry from 1998 onwards on how they should respond to developments within genetics. In the early stages, the insurance industry largely saw itself as the victim. Insurers felt that the general public, the medical profession and the biotechnology industry were accusing them of creating problems with regards to genetics. Insurers did not have an interest in genetics per se, and were not the drivers developing the technology. However, they felt obliged to respond and wanted to use genetic information if their clients had access to it. Insurers also felt that it was unfair to be blamed for public hysteria about genetics that was largely generated by the media. They also did not believe that they should be expected to educate the public about genetics.[20]

In the 1990s, the insurance industry tended to take a defensive approach, standing by the risk classification principle of 'actuarial fairness' ('each paying according to their risk') and reinforcing the consequences of adverse selection for the insurance market. These were techniques that the industry had used a decade before when it was concerned about adverse selection following HIV testing.

However, legislators in many countries did not seem to accept the argument that genetic information was similar to other medical and risk information that insurers are already allowed to request and use. In an attempt to ward off legislation that would restrict their access to genetic test results, the industry proposed various codes of practice, self-regulation and voluntary moratoria.

The industry, particularly in Europe, formed the view that, in the short-term, the actuarial impact of genetic testing was limited. The tests were not particularly predictive of future morbidity/mortality, and the numbers involved would have limited impact on their business. As Van Hoyweghen et al. pointed out, moratoria were "merely a temporary situation, and they [the insurance industry] still want to preserve the right to use genetic information in the future because they fear the prospective impact of widespread genetic testing for common diseases and its potential for adverse selection."[21] It would be easier to renegotiate a voluntary moratorium in the future than attempt to repeal legislation, and in the meantime, they could use the time to gather actuarial data on the impact for their business, lobby politicians about its consequences, and hope that the public will become less frightened about genetic technology.

6.   MORATORIA AND LEGISLATION IN EUROPE

There are restrictions on the use of genetic information within the insurance industry in most European countries.[22] Some countries have legislated to prohibit insurance companies from using genetic tests, for example, in Belgium, Denmark and France, although there have been definitional problems in what constitutes genetic information and what, in fact, is a genetic test. Voluntary moratoria or codes of practice have been adopted by the insurance industries in many other European countries. Moratoria are either indefinite (e.g. Finland, Germany), for a limited number of years (e.g. France, Ireland), or restricted to insurance policies below certain monetary values (e.g. the United Kingdom). Other countries allow the use of genetic susceptibility tests only beyond a certain level of insurability and with the consent of the individual concerned (e.g. in the Netherlands and Sweden).

From a public policy perspective, there are obvious problems with relying on self-regulatory systems if there is no external sanction imposed on a financially powerful institution, but it is a solution to drafting legislation in the rapidly developing field of genetics, where producing appropriate legal definitions is problematic.

7.   MORATORIUM IN THE UNITED KINGDOM

On 14 March 2005, a Concordat came into effect between the UK Government and the Association of British Insurers (ABI) which extended the Moratorium on insurers' use of predictive genetic tests by an extra five years, until 1 November 2011.[23] The Concordat and Moratorium will be reviewed in 2008 and updated if necessary in light of experience, research findings, developments in genetic technology, and clinical practice.

The ABI is the trade association for Britain's insurance industry, with more than 400 member companies responsible for over 97% of the insurance business in the UK. While adoption of the Concordat is voluntary, in practice it is considered to be binding on all member companies of the ABI, via its Code of Practice.

This chapter is not focused on the specific details of the Concordat, but on the language and justifications that are used, balancing private and public interest. However, in summary, the terms of the Moratorium are as follows. Customers will not be required to disclose the results of predictive genetic tests for policies up to £500 000 of life insurance, or £300 000 for critical illness insurance, or paying annual benefits of £30 000 for income protection insurance. More than 97% of policies issued in 2004 were below these limits in each category. When the cumulative value of insurance exceeds the financial limits, insurers may seek information about, and customers must disclose, tests approved by the Government-appointed Genetics and Insurance Committee (GAIC) for use for a particular insurance product, subject to the restrictions in the Concordat.

The Concordat explicitly reinforces the principle that, unless otherwise agreed, "insurance companies should have access to all relevant information to enable them to assess and price risk fairly in the interest of all their customers." Thus, applicants for life insurance should, in all normal circumstances, disclose specific risks to their health, e.g. medical information, family history or test results. The Concordat recognises the dangers of adverse selection. It explains that "[i]f the risk is not disclosed, the insurance company may face more, and more costly, claims than it was able to assume in setting the price of its insurance policies." Thus, the argument for requiring disclosure is based on fairness and justice as otherwise it could "potentially affect the future pricing or availability of insurance cover to all," so protecting other customers "from the consequences of extremely high claims, which have not been priced for."

However, the Moratorium makes an exception to this principle of disclosure by allowing patients to take a predictive genetic test without disclosing the results of that test. The majority of genetic tests confirm diagnoses of ill health and inform treatments. The Concordat is concerned only with the far smaller number of tests used to predict future illness. Patients are able to obtain 'significant' but not 'excessive' levels of coverage that might jeopardise the financial viability of the insurance industry. The Concordat suggests that insurers have been prepared to bear the risks and costs of non-disclosure because the number of policies affected by non-disclosure of predictive genetic tests is low. However, as it admits, the costs are actually "spread across the broad pool of policyholders." Thus, policyholders at population risk are expected to show solidarity with those at increased genetic risk, although of course they have not been explicitly told (nor asked for consent) that they are subsidising people in this way.

The Concordat and Moratorium claim to protect the interests of both customers and insurers, by preserving customers' access to insurance and insurers' right of equal access to information about risks. "It is designed to balance societal concerns with the need for a commercially viable, long term and fair insurance market."

8.   BALANCING PRIVATE AND PUBLIC INTEREST

Government has responsibilities in protecting the interests of the insurance industry, the general public who pay premiums and people at increased risk of developing genetic disease.

The UK Government and the insurance industry acknowledged that the Concordat was a response to concerns about the potential use of personal genetic data by insurance companies. They considered that "the relationship between medical data and insurance underwriting should be proportionate and based on sound evidence." The Concordat recognised that a minority of patients might be discouraged from taking predictive genetic tests if they fear that insurance companies may discriminate against them unfairly on the basis of the test results.

In a 1997 UK survey[24] conducted before a moratorium was in place in the UK, 28% of respondents said they would not take a genetic test if they were required to disclose the results to their insurance company.

In a statement to United States Congressional Task Force on Health Records and Genetic Privacy Preventing Genetic Discrimination in Health Insurance, Francis Collins, Director of the National Human Genome Research Institute, described the dilemma for public policy:

> "As our technology grows in genetic testing, more information will be made available to concerned individuals about their potential for developing certain conditions. While potentially providing enormous benefit by allowing individualized programs of preventive medicine, the increased availability of genetic information raises concerns about who will have access to this potentially powerful information ... Of particular concern is the fear of losing jobs or health insurance because of a genetic predisposition to a particular disease. For example, a woman who carries a genetic alteration associated with breast cancer, and who has close relatives with the disease, has an increased risk of developing breast and ovarian cancer. Knowledge of this genetic status can enable women in high-risk families, together with their health care providers, to better tailor surveillance and prevention strategies. However, because of a concern that she or her children may not be able to obtain or change health insurance coverage in the future, a woman currently in this situation may avoid or delay genetic testing ... Discrimination in health insurance, and the fear of potential discrimination, threaten both society's ability to use new genetic technologies to improve human health and the ability to conduct the very research we need to understand, treat and prevent genetic disease."[25]

Government would wish to encourage the development of genetic testing programmes that are in the public interest, for reasons outlined by Francis Collins. While there will be upfront costs in paying for the predictive tests, it is hoped that people found to be at increased genetic risk, for heart disease for example, may attempt to modify this risk by changes in behaviour and other environmental factors, e.g. by not smoking, by adopting a healthier diet, or by taking more exercise. As a consequence, treatment costs could be reduced if diseases are prevented or interventions provided at an earlier stage of the disease process. However, there is little utility in encouraging the development of such tests and establishing testing programmes if the public will not use them because of fears about how the information will be used.

In addition to the Moratorium itself, the UK Government and the Association of British Insurers therefore "agreed a set of measures intended to reassure patients so that they are not deterred from taking a predictive genetic test by fear of potential insurance consequences." These included restriction on data collection, data protection, audit of compliance, and complaint and appeals procedures.

9.   MORAL SOLIDARITY TO CARE FOR PEOPLE WITH GENETIC DISEASE

The Alzheimer's Association in the United States produced a position statement on genetic testing in 1995:

> "The presence of a gene is not a basis for underwriting insurance premiums for health care, long-term care or life insurance, nor should it be used to infringe on any individual's access to care and services."[26]

Arguably, the presence of a gene is a perfectly legitimate basis for underwriting decisions. The important thing is the accuracy of the information and how it is used. The Genetics and Insurance Advisory Committee (GAIC)[27] uses the following three criteria when examining applications for the use of predictive genetic test results in setting insurance premiums:

- **Technical Relevance**: Does the test accurately measure the genetic information?
- **Clinical Relevance**: Does a positive result in the test have likely future adverse implications for the health of the individual?
- **Actuarial Relevance**: Does a positive result justify increased premiums?

The only application to date that GAIC has approved is for Huntington's disease for life insurance policies over £500 000. All other applications have been turned down or referred back to the Association of British Insurers. The ABI has written to the Department of Health to say that it will not be submitting any applications to use predictive genetic tests, including for breast cancer, during 2006 and 2007. However, the fact that the case has been made in the context of Huntington's disease demonstrates that incorporating genetic information into underwriting decisions may be legitimate.

The Alzheimer's Association is saying something else within the second clause of this sentence of their policy statement: *"nor should it be used to infringe on any individual's access to care and services."* Here they are making a claim to a right to healthcare. No country, and certainly not the United States, gives an unlimited right to healthcare. At best, there is a guaranteed basic minimum level of health and social care provision. Instead, there may be an appeal to moral (egalitarian) solidarity, with provision of social goods according to need rather than ability to pay. An editorial in the Lancet used such an argument:

> "Many systems of healthcare provision offer treatment irrespective of particular risks or faults. This equity annoys some critics, who would like to get their money back from a climber who breaks a leg in a fall. However, the principle is worth preserving. It reflects a societal wish to provide care irrespective of circumstances ... Is there a consensus similar to that in many countries' health-service arrangements? 'We are born with our genes', it might run 'cannot alter them, and wish, as a society, that information on our genes be restricted to direct medical uses. We thereby forgo any premium advantage in being able to show that we are genetically at low risk'."[28]

Pokorski was critical of this argument:

> "In the end people will 'vote with their feet,' i.e. they will choose a solution that most closely meets their needs ... [T]he solution will almost certainly entail total acceptance of the use of genetic factors in risk classification. The explanation lies in the economic imperative that there is no viable alternative to the private insurance mechanism. If the

populace decides to endorse a system that calls for significantly higher premiums in order to subsidize others at greater risk, that is their right. For most people, however, willingness to subsidize others will quickly fade with the realization that even if they are willing to 'play by the rules' and purchase insurance 'blinded' to their genetic status, many others will not be so forthright, and the latter will use genetic information as the basis for choosing the type, amount, and timing of insurance purchases."[29]

As evidence in support of his argument, Pokorski quoted an American Council of Life Insurance survey[30] which asked insurance policyholders if they would be willing to pay more for life insurance so that everyone could receive coverage at the same rate, regardless of the risk they represented to the company. Only 27% said that they would pay more, and most wanted any increases in premiums to be limited to 10% or less, (2% of policyholders were willing to pay increases of up to 25%). This survey was not specifically about genetic risk; however, it did indicate that people are willing to demonstrate moral solidarity. Indeed, the proportion willing to pay higher premiums to exclude genetic risk from actuarial decisions may even be higher, although there is likely to be an upper limit for increased premiums that will be tolerated due to genetic adverse selection. Of course, while people may appear to be altruistic by being willing to pay higher premiums, they are actually behind a *veil of ignorance* as they probably do not know what genes they have that may modify their genetic risk of disease. Thus, they are showing solidarity on the basis that *it could be me* that will have a rejected insurance application because of a positive genetic test in the future.

Of course, governments may also be acting out of a form of *self-interest* by wanting to restrict the use of genetic information by the insurance industry. In addition to constraining healthcare costs if people who are predicted to be at increased risk attempt to modify their lifestyle risks, it is in the government's interest to encourage people to make personal provisions for health and social care. If the insurance market collapses because of the consequences of adverse selection, or people stop taking out insurance because of increased premiums, or there are excessive concerns about genetic discrimination, it is the taxpayer who will ultimately have to pick up the bill through the basic safety-net provisions for health and social care. While insurance policyholders are usually taxpayers and vice versa, the shift to more non-interventionist, laissez-faire government at the end of the twentieth century means that most governments prefer individuals to take on more responsibility for their health and social care provision, rather than make politically unpopular decisions to increase taxes.

The current compromise, as demonstrated in the voluntary moratoria between governments in Europe and the insurance industry, therefore seems advantageous to all concerned. The voluntary regulations protect:

- the public interest, by encouraging citizens to take predictive tests and to apply for insurance coverage rather than to rely on health and social care paid for out of taxation;
- the private interest of individuals, by allowing them access to a reasonable level of insurance coverage, irrespective of their gene status; and,
- the insurance industry, by preserving their market while giving them some protection from excessive claims and adverse selection, without the restrictions of legislation that can be difficult to reverse in the future.

REFERENCES

1. D. Shickle, "*On a Supposed Right to Lie [to the Public] from Benevolent Motives": Communicating Health Risks to the Public*, 3 MED. HEALTH CARE PHILOS. 241 (2000).
2. CNN.com, Genome Announcement a Milestone, But Only a Beginning, http://archives.cnn.com/2000/HEALTH/06/26/human.genome.05/index.html#r.
3. Human Genetics Commission, Public Attitudes to Human Genetic Information (2001).
4. D. W. Hadley et al., *Genetic Counseling and Testing in Families with Hereditary Nonpolyposis Colorectal Cancer*, 163 ARCHIVES INTERN. MED. 573 (2003).
5. E. A. Peterson et al., *Health Insurance and Discrimination Concerns and BRCA1/2 Testing in a Clinic Population*, 11 CANCER EPIDEMIOL. BIOMARKERS PREV. 79 (2002).

6. K. P. Geer et al., *Factors Influencing Patients' Decisions to Decline Cancer Genetic Counseling Services*, 10 J. GENET. COUNS. 25 (2001).
7. E. T. Matloff et al., *What Would You Do? Specialists' Perspectives on Cancer Genetic Testing, Prophylactic Surgery, and Insurance Discrimination*, 18 J. CLIN. ONCOL. 2484 (2000).
8. National Partnership for Women & Families on behalf of the Coalition for Genetic Fairness, Faces of Genetic Discrimination: How Genetic Discrimination Affects Real People (2004), http://www.nationalpartnership.org/portals/p3/library/GeneticDiscrimination/FacesofGeneticDiscrimination.pdf; Secretary's Advisory Committee on Genetics, Health and Society, Public Perspectives on Genetic Discrimination September 2004-November 2004 (2005), http://www4.od.nih.gov/oba/sacghs/reports/Public_Perspectives_GenDiscrim.pdf.
9. L. Low et al., *Genetic Discrimination in Life Insurance: Empirical Evidence from a Cross Sectional Survey of Genetic Support Groups in the United Kingdom*, 317 B.M. J. 1632 (1998).
10. Lloyd's, Chronology, http://www.lloyds.com/About_Us/History/Chronology.htm.
11. J. Husted, *Genetics and Solidarity, in* 1 Genetics and Insurance (T. McGleenan et al. eds., 1999).
12. *Id.*
13. Actuarial Standards Board, Risk Classification (for All Practice Areas) (ASOP number 12) (2005), http://www.actuarialstandardsboard.org/pdf/asops/asop012_101.pdf.
14. *Id.*
15. R. J. Pokorski, *A Test for the Insurance Industry*, 391 NATURE 835 (1998).
16. C. D. Zick et al., *Genetic Testing for Alzheimer's Disease and Its Impact on Insurance Purchasing Behavior*, 24 HEALTH AFF. 483 (2005).
17. Human Genetics Advisory Commission, The Implications of Genetic Testing for Insurance (1997).
18. Actuarial Standards Board, *supra* note 13.
19. I. Van Hoyweghen et al., *"Genetics Is Not the Issue": Insurers on Genetics and Life Insurance*, 24 NEW GENET. SOC. 79 (2005).
20. There is an additional problem that is worth noting. Actuaries will use genetic information as probabilistic, although there have been concerns that hitherto they have been using inaccurate predictive values within their calculations. In this sense, as far as an actuary is concerned, genetic information is no different to age, gender, smoking history. For example, women with BRCA1 mutation are more likely to develop breast cancer in a similar way as actuaries know that male drivers under the age of 25 years are more likely to be involved in a road traffic accident. In comparison, public understanding of genetics will be influenced by the media, who often use genetic information in a deterministic way i.e. if you have the gene then you will get the disease. However, the nature of genetic information and the way that it may be understood and used is beyond the scope of this chapter.
21. Van Hoyweghen, *supra* note 19.
22. H. Nys et al., Genetic Testing: Patients' Rights, Insurance and Employment. A Survey of Regulations in the European Union (2002).

23. Concordat and Moratorium on Genetics and Insurance (2005), http://www.dh.gov.uk/assetRoot/04/10/60/50/04106050.pdf.
24. E. J. Brunner et al., *Public Is Concerned About Gene Testing*, 314 B. M. J. 1552 (1997).
25. Collins F. Statement to The Congressional Task Force on Health Records and Genetic Privacy Preventing Genetic Discrimination in Health Insurance (1997), http://www.genome.gov/10002352.
26. Alzheimer's Association, Position Statement on Genetic Testing (1995), http://www.alz.org/Advocacy/downloads/statements_genetictesting.pdf.
27· Genetics and Insurance Committee, Fourth Report from January 2005 to December 2005 (2006).
28. Editorial, *Have You Had a Gene Test?* 347 LANCET 133 (1996).
29. R. J. Pokorski, *Insurance Underwriting in the Genetic Era*, 80 CANCER (supplement) 587 (1997).
30. M. E. Dixon et al., Monitoring Attitudes of the Public (1995).

Public Opinion, Consent and Population Genetic Biobanks

Timothy CAULFIELD
Nola M. RIES[1]

Health Law Institute, University of Alberta, Canada

1. INTRODUCTION

With an increasing number of large-scale population genetic biobanks emerging throughout the world, the issue of how best to obtain consent from individual participants has become a major policy concern. This is because traditional consent and research ethics norms typically require consent for each new use of identifiable health information. However, some argue this requirement stifles important research and that alternate consent models, such as a one-time agreement to participate in future research, are preferable and are acceptable to the public.

For example, David Wendler published an editorial in the *British Medical Journal* in March 2006 suggesting that the available opinion data "provide compelling evidence that one-time general consent is the best option."[2] In another commentary, Mark Rothstein argues that "[a]lthough some individuals and groups adhere to the position that blanket consent for future research is permissible only when the samples are anonymized, this position is unnecessarily paternalistic and threatens to impair research. As long as the potential research subjects are clearly apprised of the range of possible future uses of their sample, they should be permitted to give one-time blanket consent to such uses."[3] Other analysts have reached similar conclusions.[4]

While survey research to which authors like Wendler refer is valuable and helps inform biobank policy, it is far from definitive. Opinion surveys are subject to a high degree of interpretation and have methodological shortcomings.[5] Most importantly, public opinion may be either more lenient or more conservative than existing legal and ethical requirements, and researchers must ensure consent policies comply with those rules. The challenge of how best to obtain consent without unduly hindering research must be resolved having regard not only to measures of what the public would consider acceptable, but also to fundamental principles that underlie legal and ethical rules.

In this chapter, we revisit the issue of public opinion, highlighting data that suggests individuals are not uniformly in favour of one-time consent models for biobanks. We also comment on legal and ethical norms that underpin the regulation of population genetic research and identify tensions between compliance with informed consent requirements and research interests in biological samples and personal information. We note examples from several jurisdictions where consent laws have been amended to address concerns that specific informed consent rules hindered research and other legitimate uses of information. We argue that legal clarity regarding consent rules is critical for valuable population genetic research to proceed, but caution that public opinion is just one piece of the puzzle to guide regulation in this field.

2.   REVISITING PUBLIC OPINION

If one examines the studies Wendler uses to support his contention that "one-time general consent is the best option," it is, in fact, difficult to understand how he could arrive at such a strong conclusion. The data from the studies referenced in the Wendler article do not paint a coherent picture of public opinion. For example, one of the primary conclusions of the Goodson study, which Wendler specifically uses to support his thesis, is that the data "demonstrates that a relatively large number of individuals would want *ongoing* control over their tissue."[6] Likewise, some of the other studies found that a relatively small majority felt comfortable with a blanket consent approach, even in countries where one would expect a higher level of support. For example, in the Hoeyer et al review of the Swedish study of the general public, 66.8% supported the contention that "there are occasions when an REC [research ethics committee] may decide."[7] Approximately eighty percent (79.4%) of *these* individuals felt that a "previously provided permission to use the tissue in other research projects" was a situation when they approve of a research ethics board providing consent on their behalf. In other words, just over half of the total felt comfortable with a blanket consent approach – hardly a ringing endorsement, particularly from a country with a strong community-oriented ethos and a long history of cohort studies. This same study found that 48% "feel respected and involved" by "repetitive informed consent procedures" and only 11.3% feel they received "superfluous information."

Other studies relied on to support the blanket consent approach reported on consent experiences of research participants in projects where fresh consent was not an option[8] and did not study their preferences for consent models more generally. In other words, many of these studies simply asked about the acceptability of an existing approach. While useful, this data does not reflect what the public views as the best approach.

Moreover, numerous other studies have come to an opposite conclusion regarding blanket consent. In some respects, this is because these studies explored public preference, rather than just the acceptability of a blanket consent approach. For example, in one study where various consent scenarios were proposed in the context of UK Biobank, it was found that the most highly preferred scenario was "consent every time new data is required...."[9] Likewise, a UK Human Genetics Commission study found that 82% of the respondents either strongly agree (44%) or tend to agree (38%) that fresh consent must be sought.[10]

Other research demonstrates that many believe genetic information is special and worthy of more stringent protection than other forms of health information. For example, a Canadian study found that 90%

either strongly agree (61%) or agree (29%) that genetic information is different from other personal information and rules governing access should be stricter.[11] A Japanese study compared attitudes of participants in a non-genetic cohort study with those in a genetic cohort study and found that "the general population responds skeptically towards participation in genetic research when actually faced with the decision-making process."[12] The UNESCO *International Declaration on Human Genetic Data* formally adopts the view that genetic information is unique and states that "special protection should be afforded to human genetic data and to biological samples."[13]

Our different interpretation of available public opinion data does not suggest we believe Wendler has misrepresented it. On the contrary, our divergent reading of the data illustrates the challenges in attempting to discern public views and to determine how those views ought to influence the development of consent policies for population genetic biobanks. Clearly, "decisionmaking in a democratic society should take account of public attitudes, and, therefore, public engagement must be central in planning for and implementing a large population project."[14] However, public opinion data—whatever it reveals—cannot supplant existing legal and ethical rules.[15]

3.   LEGAL AND ETHICAL NORMS REGARDING PARTICIPATION IN POPULATION GENETIC RESEARCH

Absent explicit legislation to the contrary, prevailing legal and ethical norms are generally interpreted as requiring specific consent for each new use of identifiable personal information, with limited exceptions that will permit a more general form of consent. For example, in jurisdictions with a stringent legal standard requiring disclosure of full information about the nature of a research study and its risks (no matter how remote or rare), fresh consent for each new research project is a logical interpretation of existing consent law.[16] Ethical norms also emphasize specific informed consent[17] and the World Health Organization has stated that: "Blanket consent for future research is only permissible in circumstances where anonymity of future data can be guaranteed. ... If this guarantee is not possible, or if linking of data is necessary for the research, then specific consent to the specific research must be obtained."[18] Current commentary increasingly emphasizes the capacity to re-identify data, so "guaranteeing complete confidentiality may never be possible."[19]

The legal and ethical emphasis on individual autonomy implies that reasons to justify deviating from specific consent must be pressing and substantial. Concern about consent bias, cost or convenience must be significant to support erosion of a fundamental right. It is worth considering whether research inconvenience could ever stand as a legitimate justification to erode an individual's right to consent, though some researchers argue that legally-mandated consent requirements go beyond inconvenience to seriously hinder the feasibility of research involving personal information and/or biological samples.[20]

However, the right to give consent to participate in research is an extension of the human right to make decisions about things that are intimate to personhood.[21] Public opinion may also be skeptical of waiving specific consent in favour of advancing research interests. In the Hoeyer study, only 37.6% of the 66.7% who approved of surrogate consent by a research ethics committee (REC) thought cost was an acceptable justification. The alleged scientific value of the research also does not necessarily trump specific consent. The right of self-determination, particularly in the area of biomedical research, has grown substantially post-World War II. Indeed, the paramountcy of autonomy in health law jurisprudence and research ethics policy is a reaction against the "public worth of science" argument.

This is not to say that the right of autonomy is absolute. There are many examples where the state, for the good of the public, overrides individual autonomy, such as in areas of public and mental health. Individuals with communicable diseases may be detained for treatment against their will[22] and mentally ill persons who pose harm to themselves or others may be committed without consent.[23] However, these situations provide more compelling justification for limiting autonomous decision-making than the research context as community members are more likely to face imminent harm in public and mental health situations. As well, in these contexts, legislators have enacted detailed legal regimes to stipulate when individual autonomy may be overridden and to specify safeguards to ensure individual rights are restricted only to the degree necessary to achieve legitimate health goals.

In regard to research use of biological samples, some argue strongly that the ethical imperative of solidarity, which is concerned with helping others, should override individual rights to control biological samples, especially when those samples cannot identify an individual. At the extreme, it is argued that individuals have a moral duty to participate in health research because they benefit from knowledge that results from research and refusal to participate is a form of free ridership that deprives future generations of improved knowledge. John Harris maintains that "almost everyone now living, certainly everyone born in high income industrialised societies, has benefited from the fruits of past research"[24] and, consequently, bears some obligation to participate in research. He notes that a person may hesitate to consent to research with biological samples because "I may feel that since I understand little of the future uses for my tissue it would be safer to say 'no'."[25] Harris gives little weight to this position, arguing that it is short-sighted and fails to take account of all factors that are relevant to weighing rights and interests of (potential) research subjects and those who benefit from research.

Some ethicists also contend that a portion of research participants favour a one-time consent and, in these cases,

> "it is unethical to burden research participants unnecessarily with more information than they want. ... It is not necessary ... to give renewed information or to contact these participants in order to check whether they feel they need more information .... We owe them respect as moral agents regardless of whether they want as much information as possible or whether they settle for a minimum."[26]

In weighing the cost and benefit of elaborate informed consent procedures for secondary research, some argue that complicated paperwork does "not recognise personal autonomy – the right to be able to agree with the proposal without the unnecessary administrative burden on the participants and research team."[27] It may further be argued that if we truly respect individual autonomy, we should respect an individual choice to participate in population genetic research on the basis of a one-time general consent process. As Harris states, "it is usually the best policy to let people define and determine 'their own interests'."[28]

However, views supporting general consent procedures in place of specific informed consent are not uniformly accepted, and while ethical arguments may be advanced in favour of participation in biomedical research, the law typically does not conscript individuals into research,[29] nor does the law generally permit waiver of specific informed consent unless certain criteria are met. For instance, some Canadian personal information protection laws authorize waiver if consent from individuals is not practicable, a research ethics board authorizes waiver, the anticipated benefit of research is expected to outweigh any harms (such as harm to personal privacy), and appropriate security safeguards are in place.[30] Further, legal regimes vary from jurisdiction to jurisdiction, resulting "in considerable variation in the domestic law that applies to the use of DNA samples, personal information and medical records...."[31]

4.   TENSIONS AND LESSONS LEARNED

Many recommendations for handling consent in population genetic biobanks appear to be at odds with existing legal requirements (or at least with how existing legal principles may be interpreted). This leads to uncertainty for researchers in how to proceed with establishing consent and governance frameworks for these projects. This problem deserves legislative attention, and it is worth considering how the debate over consent for collection, use and disclosure of personal information has played out in a variety of health contexts, including registries, electronic health records and biobanks. Lessons learned may be instructive in determining how to clarify legal rules for consent to participate in population genetic biobanks.

Interestingly, a trend appears where legislation is enacted to establish stringent consent requirements for use of personal health information, then is amended after adverse consequences arise. For example, in the 1980s, legislation governing cancer registries in two German states was amended to require informed consent for inclusion of information in the registry. Following this legislative change, the number of cancer cases reported to the registries dropped by over 70%, which seriously compromised the registries' value. The legislation was subsequently amended to relax the consent requirement.[32] Debate about consent models has also arisen in the context of collecting and sharing patient information via electronic health records. One Canadian province required specific patient consent before information could be disclosed electronically. However, the government removed this statutory provision based on operational challenges of obtaining patient consent.[33]

Similarly, in 2004, the New Zealand government amended the national *Code of Health and Disability Services Consumers' Rights*[34] to relax consent requirements regarding use of biological samples for research. Before the amendment, the Code stipulated that a sample could only be used for subsequent research with the individual's informed consent. Following criticism that this requirement was too onerous, the Code was revised to remove the informed consent requirement if the research has received ethical approval or is used for quality assurance purposes.[35]

UK commentators argue that the *Data Protection Act 1998* "permits the proper use of personal information without always seeking informed consent."[36] Yet, problems arise in practice when "those who control access to healthcare data are not allowing these legitimate and sensible exceptions to be put into practice."

The Icelandic experience with establishing a national health database, with linkages among healthcare and genealogical records and genetic information derived from donated biological samples, highlights the challenges in formulating legislation for population genetic biobanks. One summary offers the following description:

"After a vigorous debate in **Icelandic** society, the **Icelandic** parliament passed a law permitting the construction of the IHD [Icelandic Healthcare Database], a data base made up of information from the medical records of all **Icelandic** citizens. The debate included 700 newspaper articles, more than 100 radio and television programs, and several town meetings all across Iceland. On the eve of the parliamentary vote, a poll showed that 75 percent of Icelanders supported the passage of the bill, whereas 25 percent were against it. The data-base law was passed by the same margin, and since then support for it has been growing. A poll taken by the Gallup organization in the beginning of April 2000 showed that 90 percent of those who took a stand on the issue supported the data-base law, and 10 percent were against it."[37]

However, the Icelandic law, which establishes a presumed consent model for the IHD, has been successfully challenged for infringing constitutional privacy protections[38] and over 20 000 citizens (out of a total of 270 000) have chosen to opt out of participating in the database.[39]

5.   MOVING FORWARD

Many scientists believe population genetic biobanks and cohort studies are an absolutely essential research tool – one of the best ways to tease out the complex role of genes and the environment in the development of disease. As such, devising consent models that will allow this research to move forward is a worthy goal. At the same time, legal and ethical rules must achieve an appropriate balance of pertinent rights and interests.

The public's willingness to participate in research depends on trust in researchers. Some public opinion data suggests that because individuals trust researchers, they are willing to donate samples to biobanks.[40] Because this trust relationship exists, it is further argued, one-time general consent is sufficient because individuals have faith that researchers will use their personal information appropriately. Paradoxically, the failure of researchers to comply with legal and ethical rules may undermine trust, especially when scandals about misuse of human tissues emerge.[41]

For population genetic research to move forward on unambiguous legal and ethical terrain, clear rules are required. Existing laws and ethics guidelines regarding collection, use and disclosure of personal information often impose requirements to obtain specific informed consent. We have highlighted some examples where legislative regimes have been amended to attenuate consent requirements after negative consequences arose. A body of research is growing to demonstrate potential barriers that existing laws have on research-related initiatives. This data, combined with results of public opinion survey, form a basis for legislators to engage the public more actively[42] and evaluate how best to regulate population genetic biobanks. Amending existing laws or developing new legislation to clarify rules for population genetic research is a complex undertaking and consensus on ideal consent approaches is unlikely. Nonetheless, everyone in society has a stake in this growing area of scientific inquiry, and public engagement is a critical step to moving forward.

REFERENCES

1. Timothy Caulfield is Research Director, Health Law Institute, and Professor, Faculty of Law and Faculty of Medicine & Dentistry, University of Alberta, and Canada Research Chair in Health Law & Policy. Nola M. Ries is Research Associate, Health Law Institute, University of Alberta and Adjunct Assistant Professor, Faculty of Human & Social Development, University of Victoria. The authors thank Genome Alberta for funding support and are very grateful to Alethea Adair for her excellent research and editing assistance and for her comments on a draft of this chapter.

2. D. Wendler, *One-time General Consent for Research on Biological Samples*, 332 BRIT. MED. J. 544 (2006).

3. M. A. Rothstein, *Expanding the Ethical Analysis of Biobanks*, 33 J. LAW. MED. & ETHICS 22 (2005).

4. *See, e.g.,* M. G Hansson et al., *Should Donors be Allowed to Give Broad Consent to Future Biobank Research?* 7 LANCET ONCOL. 266, 269 (2006): "We suggest that consent should be regarded as valid until further notice. There should be a realistic opportunity for withdrawal of consent for those who have donated identifiable samples and data, which can be an issue with biobank-based research that can last for many decades (eg, contact information to the biobank might have changed, and donors are unlikely to save contact information for decades)."

5. M. Nisbet, *The Polls—Trends: Public Opinion About Stem Cell Research and Human Cloning*, 68 PUBLIC OPIN. Q. 131 (2004).

6. M. L. Goodson & B. G. Vernon, *A Study of Public Opinion on the Use of Tissue Samples from Living Subjects for Clinical Research*, 57 J. CLIN. PATHOL. 135 (2004).

7. K. Hoeyer et al., *Informed Consent and Biobanks: A Population-Based Study of Attitudes Towards Tissue Donation for Genetic Research*, 32 SCAND. J. PUB. HEALTH 224 (2004).

8. G. M. McQuillan et al., *Consent for Genetic Research in a General Population: The NHANES Experience*, 5 GENET. MED. 35 (2003). *Also* K. Hoeyer et al., *The Ethics of Research Using Biobanks: Reason to Question the Importance Attributed to Informed Consent*, 165 ARCH. INTERNAL MED. 97 (2005). In this study, a survey was sent to a random sample of people participating in Sweden's Medical Biobank. The survey "asked whether respondents knew that they could withdraw their informed consent to donate: of 918 respondents, 290 (31.6%) answered positively, 511 (55.7%) answered negatively, and 117 (12.7%) were not aware that they had consented" (98). Another study Wendler uses is D.T. Chen et al., *Research with Stored Biological Samples: What do Research Participants Want?* 165 ARCH. INTERNAL MED. 652 (2005), which analyses consent forms for research participants enrolled in clinical research at Warren G. Magnussun Clinical Centre, National Institute of Health.

9. R. Hapgood et al., *Public Preferences for Participation in a Large DNA Cohort Study: A Discrete Choice Experiment*, *(Sheffield Health Economics Group, Discussion Paper Series)* (2004).

10. UK Human Genetics Commission, *Public Attitudes to Human Genetic Information* (MORI Social Research, 2000).

11. Pollara & Earnscliffe Res. & Comm., Public Opinion Research Into Biotechnology Issues Third Wave 51 (2000).

12. K. Matsui et al., *Informed Consent, Participation In, and Withdrawal From Population Based Cohort Study Involving Genetic Analysis*, 31 J. MED. ETHICS 385 (2005).

13. *International Declaration on Human Genetic Data*, G.A. Res. 22, at 39, UNESCO 32nd General Conference, 32 C/Res. 22 (Oct. 16, 2003). Additionally, willingness to consent to donation of biological samples for research may vary depending on the type of sample requested. Parents may hesitate to consent to donation of their children's samples and some samples may be perceived as more intimate or personal than others. A U.S. study of female participants in a maternal and child cohort study (J.L. Daniels et al., *Attitudes Towards Participation in a Pregnancy and Child Cohorts*, 20 PEDIATRIC & PERINATAL EPIDEMIOLOGY 260 (2006)) examined willingness to provide consent for vaginal swabs, blood sample and saliva collection. Interestingly, the study reports that "[p]roviding saliva was most troublesome to women, even though other procedures might usually be considered more invasive." (p. 263). However, many marked 'maybe' or 'no' when asked about their child participating in developmental assessments (26% maybe, 10% no), swabbing the child's cheek to assess nutritional status (21% maybe, 13% no) or collect DNA (33% maybe, 22% no), or providing a blood sample (40% maybe, 34% no)" (pp. 263–64).

14. Secretary's Advisory Committee on Genetics, Health, and Society, Policy Issues Associated with Undertaking a Large U.S. Population Cohort Project on Genes, Environment and Disease, A Draft Report 21 (2006).

15. But public opinion data certainly may indicate areas where the public supports revising those rules.

16. H. T. Greely, *Breaking the Stalemate: A Prospective Regulatory Framework for Unforeseen Research Uses of Human Tissue Samples and Health Information*, 34 WAKE FOREST L. REV. 737 (1999).

17. Canadian Institutes of Health Research, Natural Sciences and Engineering Research Council of Canada, Social Sciences and Humanities Research Council of Canada, *Tri-Council Policy Statement: Ethical Conduct for Research Involving Humans (*1998, with 2000, 2002 and 2005 amendments). *See* especially Article 2.4: "Researchers shall provide, to prospective subjects or authorized third parties, full and frank disclosure of all information relevant to free and informed consent."

18. *International Declaration on Human Genetic Data, supra* note 13.

19. I. S. Kohane & R. B. Altman, *Health Information Altruists—A Potentially Critical Resource*, 119 NEW ENG. J. MED. 2074, 2075 (2005).

20. *See, e.g.*, J. V. Tu et al., *Impracticability of Informed Consent in the Registry of the Canadian Stroke Network*, 250 NEW ENG. J. MED. 1414 (2004); L. Trevena et al., *Impact of Privacy Legislation on the Number and Characteristics of People who are Recruited for Research: A Randomized Controlled Trial*, 32 J. MED. ETHICS 473 (2006).

21. V. Arnason, *Coding and Consent: Moral Challenges of the Database Project in Iceland*, 18 BIOETHICS 27 (2004).

22. *See, e.g.,* Toronto (City, Medical Officer of Health) v. Deakin, [2002] O.J. 2777 QUICKLAW (O.C.J. July 3, 2002).

23. *See, e.g.,* Fleming v. Reid, [1991] 4 O.R. 3d 74. However, in the mental health context, one could argue that a person who no longer has capacity to give consent to treatment because of mental illness is not an autonomous decision-maker.

24. J. Harris, *Scientific Research is a Moral Duty*, 31 J. MED. ETHICS 242, 243 (2005).

25. *Id.*

26. S. Eriksson & G. Helgesson, *Keep People Informed or Leave Them Alone? A Suggested Tool for Identifying Research Participants who Rightly Want only Limited Information*, 31 J. MED. ETHICS 674 (2005).

27. P. Singleton & M. Wadsworth, *Consent for the Use of Personal Medical Data in Research*, 333 B. M. J. 255, 255 (2006).

28. Harris, *supra* note 24, at 244.

29. An exception to this statement exists where laws compel centralized reporting and collection of health information into repositories such as disease registries. The impetus for mandatory reporting is typically for health surveillance activities, but registry information is often used for research purposes.

30. For discussion of Canadian personal information protection legislation in the context of population genetic research, *see* T. Caulfield & N. M. Ries, *Consent, Privacy and Confidentiality in Longitudinal, Population Health Research: The Canadian Legal Context*, HEALTH LAW J. (supplement) 1 (2004).

31. J. Kaye, *Do We Need a Uniform Regulatory System for Biobanks Across Europe?*, 14 EURO. J. HUM. GEN. 245, 245 (2006).

32. J. R. Ingelfinger & J. M. Drazen, *Registry Research and Medical Privacy*, 250 NEW ENG. J. MED. 1542 (2004).

33. *See* N. M. Ries, *Patient Privacy in a Wired (and Wireless) World: Approaches to Consent in the Context of Electronic Health Records*, 43 ALBERTA L. REV. 681 (2006).

34. Code of Health and Disability Services Consumers' Rights 1996, 1996 S.N.Z., http://www.hdc.org.nz/act_code/the_right_of_code/thecode.html.

35. For the Minister of Health's explanation for the legislative amendment, *see* http://www.hdc.org.nz/files/files/web_review_of_act&code_june_2004.pdf.

36. Data Protection Act 1998, 1998, c. 29 (Eng.); A. Iversen et al., *Consent, Confidentiality, and the Data Protection Act*, 332 BRIT. MED. J. 165, 166 (2006).

37. J. R. Gulcher & K. Stefánsson, *The Icelandic Healthcare Database and Informed Consent*, 342 NEW ENG. J. MED. 1827 (2000).

38. Ragnhildur Guomundsdottir v. The State of Iceland, [2003] 151 Icelandic Supreme Court, http://www.epic.org/privacy/genetic/iceland_decision.pdf.

39. A. Cambon-Thomsen, *The Social and Ethical Issues of Post-Genomic Human Biobanks*, 5 NAT. REV. GEN. 866 (2004).

40. In addition to studies Wendler cites, *see* also A. Kettis-Lindblad et al., *Genetic Research and Donation of Tissue Samples to Biobanks. What do Potential Sample Donors in the Swedish General Public Think?*, 16 EURO. J. PUB. HEALTH 433 (2006). The authors observe that "[a]ttitudes towards genetic research and trust in authorities seems to be crucial for the willingness to donate...."

41. A notable example is the Alder Hey scandal in the UK involving unauthorised removal, retention, and disposal of children's organs and other tissues at a children's hospital from 1988 to 1995. For background, *see, e.g.,* M. Hunter, *Medical Research under Threat after Alder Hey Scandal*, 322 BRIT. MED. J. 448 (2001).

42. For discussion of differences between public engagement and public consultation, *see, e.g.,* D. Castle & K. Culver, *Public Engagement, Public Consultation, Innovation and the Market*, 6 THE INTEGRATED ASSESSMENT J. 137 (2006).

Challenges for Public Health Genomics – the Public Health Perspective on Genome-based Knowledge and Technologies

Angela BRAND

German Center for Public Health Genomics (DZPHG), University of Applied Science, Bielefeld, Germany

1. BACKGROUND

Medicine is currently undergoing an extraordinary development from its morphological, phenotypic orientation to a molecular, genotypic orientation,[1] thus promoting the importance of prognosis and prediction.[2] What about public health?

To date, public health practice has concerned itself with environmental determinants of health and disease and has paid scant attention to genetic variations within the population or between populations. The advances brought about by genomics are changing these perceptions. Many predict that this knowledge will enable not only clinical interventions but also health promotion messages and disease prevention programmes to be specifically directed or targeted to susceptible individuals or to subgroups of the population, based on their genetic profile and risk stratification. Obviously, the integration of genome-based knowledge and technologies into public health research, policies and health services for the benefit of all will be one of the most important future challenges that our healthcare systems will face.

"...It is clear, that the science of genomics holds tremendous potential for improving health globally.... The specific challenge is how to harness this knowledge and have it contribute to health equity, especially among developing nations...".

This quotation by Gro Harlem Brundtland, former Director General of the World Health Organization (WHO), is found in the 2000 "Report of the Advisory Committee on Health Research." Craig Venter, former president of Celera Genomics, also stressed the significance of this issue at a symposium about the future of public health at the Harvard School of Public Health:

"Three years ago the human genome – the "book of life" – was largely unknown. Today, anyone can read what it contains. Genomics is already providing fascinating insights into our species' evolution and clues to the some of the differences between individuals in susceptibility of diseases. The key question for public health, however, is whether it will improve the health of all of the world's people, or whether it will just widen the technology gap between rich and poor. Ask people what they understand of the potential of genomics for human health, and many will talk about an unprecedented opportunity to develop new drugs and vaccines. Others are concerned that the poor will gain nothing, while the rich will gain a kind of "boutique medicine": the opportunity to buy a full analysis of their personal genetic makeup, and then purchase designer therapies. If genomics is to make a major impact on global health, it will have to help provide affordable population-wide tools for combating common diseases...".

Of course, there are compelling reasons to think globally in terms of global health and genomics,[3] but first, one has to act locally. The key question is whether "the right things" are done on local level: are the current public health strategies evidence-based, i.e. do we assure the "right" health interventions (concepts of health needs assessment (HNA) and health technology assessment (HTA)) in the "right" way (concepts of quality management and policy impact assessment (PIA)) in the "right" order and at the "right" time (concepts of priority setting and health targets) in the "right" place (concepts of integrated healthcare and health management)? Since so far there has been almost no integration of genome-based knowledge and technologies into all of these concepts, it becomes clear that current public health strategies are not evidence-based at all. Thus, the public health agenda demands a vision that reaches beyond the research horizon to arrive at application and public health impact.[4] What is the role of genomics in this scenario?

2.   THE CHALLENGES FOR PUBLIC HEALTH

European and US public health institutions and platforms like the Public Health Genetics Unit (PHGU) in Cambridge, UK, the German Center for Public Health Genomics (DZPHG) in Bielefeld and the US Office of Genomics and Disease Prevention at the Centers for Disease Control and Prevention (CDC) in Atlanta, which work closely with researchers from genetic and molecular science ("modern biology") as well as from population science, the humanities and social science, are much more optimistic and clear about the relevance of genomics for public health than are others.[5] Interestingly, they all have strong links or are even part of their respective national genome research projects and are translating genome-based knowledge from biotechnology and biobanks through genetic epidemiology into public health ("translational research"). By using methods like horizon scanning, fact finding and monitoring to identify research trends as early as possible, they are already doing a prospective evidence-based evaluation, i.e. an evaluation that is carried out in the process of basic research and not just in the (retrospective) process of implementing public health strategies and policies,[6] which will always tend to lag behind.

In the last twenty years, advances in genome-based research have revolutionized knowledge about the role of inheritance in health and disease.[7] In the past, there was a narrow focus that looked only at the role of inheritance in monogenetic diseases (the human genetics setting). At present, the role of genetic susceptibilities and other biomarkers in complex diseases is already discussed (the medical, community health setting as well as the public health setting). But in the future, the focus will be even broader, analysing the role of genetic determinants together with other health determinants in health problems (the public health setting). For example, it is now known that DNA determines not only the cause of single-gene disorders, which affect millions of people worldwide, but also predispositions ("susceptibilities"),[8] which are based on genotype and haplotype variants,[9] to common diseases. New technologies will allow researchers to examine genetic mutations at the functional genomic unit level[10] and to better understand the significance of environmental factors such as chemical agents, nutrition or personal behaviour[11] in relation to the causation of diseases like cardiovascular diseases,[12] allergies, cancer, psychiatric disorders or infectious diseases.[13]

Evidently, these rapid advances in genomics and its accompanying technologies are triggering a shift in the comprehension of health and disease as well as in the understanding of new approaches to prevention and therapy.[14] Which consequences can be drawn from this knowledge, and how can it be translated into policy[15] and practice in a responsible and timely manner?

Clarifying the general conditions under which genome-based knowledge and technologies can be put to best practice in the field of public health, paying particular consideration to the public health-specific ethical, legal and social implications (ELSI),[16] are currently the most pressing tasks in the emerging field within public health, variously defined as public health genetics or public health genomics (PHG). As it aims to apply genetic and molecular science to the promotion of health and disease prevention through the organised efforts of society, integral to its activities is dialogue with all stakeholders in society, including industry, governments, health professionals and the general public.[17] Thus, the integration of genome-based knowledge and technologies into public health research, policy and practice will be one of the most important future challenges for all healthcare systems.[18] Expertise is already available and can be clustered and evaluated for socially accountable use.

For example, in a condition like coronary heart disease, to be a heterozygote for the LDL receptor gene confers an increased risk for developing the condition. But, as is also true for all other risk factors (e.g., social factors, diet, smoking, physical activity) which have been identified by epidemiologists in the past few decades, the presence of the genetic biomarker is not predictive: those who have it may not develop the disease, while those without it may end up with the disease.[19] Obviously, the scenario is very much like that of coronary heart disease in the presence of raised blood pressure or cholesterol levels: the increased risk implies "only" a (high or low) probability, and the genetic biomarker is "just" another modifier in the causality of the disease and therefore exceptional.[20] Nevertheless, the ethical question is how we will handle these susceptibilities. As a first step to answering this question, large-scale population-based epidemiologic studies are needed to measure associations between specific gene variants and environmental factors and the risk of coronary heart disease.[21] For translating such discoveries into interventions, it is necessary not only to quantify the impact of gene variations on risk for the condition, but also to quantify the effect of modifiable factors that interact with gene variations.[22] Based on the knowledge of these attributable risks, sound policies and effective interventions can be developed.[23] Regarding infectious diseases, research is being expanded to include family histories and host genetic factors that influence susceptibility to or severity of certain infectious diseases and that also affect responsiveness to vaccines and therapies. The identification of several gene-disease associations for parasitic (e.g., malaria), viral (e.g., HIV or hepatitis) and bacterial (e.g., tuberculosis or cholera) infections provide critical clues to control these infectious diseases. In this way, public health strategies will be more effective and efficient.

Policymakers must be aware of the current challenge to improve consumer protection; to monitor the implications of genome-based knowledge and technologies for health, social and environmental policy goals; and to assure that genomic advances will be tailored not only to treat medical conditions, but also to prevent disease and improve health.[24] Sound and well reflected genetic policies and programmes require a timely and coordinated process of evidence-based policymaking that relies on scientific research and ongoing community consultation.[25] An acceptable and maybe delicate balance between providing strong protection of individuals' interests[26] and, at the same time, enabling society to benefit from the genomic advances must be found.[27]

Identifying needs of genetic tests as well as of genome-based information and technologies[28] (e.g., by using the method of Health Needs Assessment (HNA)), weighing the benefits and risks of predictive genetic tests and genetic screening interventions[29] (e.g., by using well established public health methods such as Health Technology Assessment (HTA)), assessing the benefits of preventive strategies and analysing complex new problems such as "genetic inequalities"[30] and genome-based technologies such as microarrays, are essential. On the one hand, even if in terms of genetic susceptibilities and polymorphisms, it will turn out that "we are all at risk for something," there is potential for social inequalities in health as well as for social exclusion: if genetic tests will not be covered by sickness funds, there will be a two-tier system for access to genome-based knowledge and thus to individualized and stratified prevention, diagnostics and therapy. On the other hand, even if genetic tests will be reimbursed in most healthcare systems, as should be the case, there will be another ethical and social problem that may be much more discriminatory: since genomics comprises extremely complex information, public health professionals will have the task of empowering and enabling people not only to understand this novel knowledge, but also to make people capable of sound decision-making regarding the application of genetic tests[31] and genome-based information, and therefore to assure a fair equality of opportunities. Otherwise, the gap between people who are able to handle this complexity and those who are not will have the potential to create a new kind of social inequality.[32] For the future, this supports a conception of public health taking leadership by implementing an evidence-based model of policymaking. This is the reason why in the US, the UK and Germany, public health genomics has been seen as the integration of genome-based knowledge and technologies into public health research, policy and practice for the benefit of population health.

For the public health community, it is important to stress that public health genomics has nothing to do with modifying genes and that "genetic determinism" and "genetic exceptionalism" are obsolete.[33] In addition, it must be clarified that public health genomics is not synonymous with genetic epidemiology in the same way public health is not synonymous with epidemiology; also, community genetics[34] is not synonymous with public health genomics as community health is not synonymous with public health.[35] Medicine (and here mainly human genetics), community genetics and public health genomics can be understood as complementary. While in the medical setting the focus is on the use of genetic tests and other biomarkers in clinical practice, in public health genomics the focus is on the use of genetic determinants together with other health determinants in the healthcare system. Community genetics is the bridge between both settings. Furthermore, in terms of public health genomics, the idea of integrating genome-based knowledge and technologies into the aims and tasks of public health should be understood and promoted.

3. PUBLIC HEALTH ISSUES AND PRIORITIES

During the past century, achievements in public health led to enormous improvements and benefits in the health and life expectancy of people around the world. Immunization programmes and better sanitation practices resulted in the eradication or reduction of many infectious diseases as well as in safer food and water supplies. Advances in occupational safety considerably decreased the number of work-related injuries, illnesses and deaths. In the past 30 years, identification of behavioural risk factors, such as smoking, inactivity and poor dietary habits, gave rise to educational interventions and a decline in death rates from certain chronic diseases.

As for future achievements in public health, the CDC Office of Genomics and Disease Prevention predicts: "Perhaps because of these accomplishments, the determinants of disease and disability— whether natural or human made—are often perceived as originating outside the body. Although it has long been recognized that disease generally results from a constellation of host- and environment-specific factors, scientific and technologic limits have concentrated attention on the environment. Exogenous influences will continue to be vital for public health, but focusing solely on these influences may lead to diminishing rates of return compared to the triumphs of the past. To continue making significant strides, the effectiveness of public health interventions must be strengthened by more fully incorporating knowledge of internal, host-specific factors and their interactions with environmental exposures including the social environment and lifestyles…"

In the realm of social policymaking, there is a need to come up with a clear strategy for assessing and translating this novel knowledge and application in real time. Policymakers now have the opportunity to take action. A precondition for immediate action is strategic planning across health programmes, promoting genomic competency among all health professionals, enhancing surveillance and epidemiologic capacity (e.g., by combining already existing DNA-based biobanks and integrating them into well-established surveillance systems) to support evidence-based policymaking, building partnerships and, finally, seeking input from stakeholders. Integrating genome-based information into health communication will be an essential tool to generate distributed knowledge.

Likely benefits as well as potential risks of the integration of genomics into public health interventions (assessment) should be identified. The framework (corridors) for effective, efficient and socially acceptable policies (policy development) should be described. And steps and ways should be proposed to assure these policies are used in public health practice (assurance). At the same time, these three steps ("public health trios") describe the core functions of public health agencies at all levels of government.[36]

One specific task of public health genomics is to rethink and systematically evaluate every condition of interest to public health.[37] There is the potential for much more target-oriented and stratified prevention strategies[38] to ultimately replace "one strategy for all." Moreover, there is clearly potential to avoid ineffective or even "faulty" preventive strategies. For example, there is already the challenge of differentiating between individuals who respond to certain vaccinations and those who do not. Why, then, should non-responders take the risk of side-effects from vaccination if the vaccination will be ineffective and will have no benefit to them? In this specific situation, which is estimated to be true for at least 10% of the population, would not this kind of primary prevention be immoral? As another example, obesity is not only influenced by lifestyle habits such as inactivity and nutrition, but also (in more than 60%) by several genetic factors. Furthermore, it is triggered by many other factors, such as infectious diseases and social factors. In at least 2% of these 60% of cases, obesity is only due to mutations in the MC4R-gene. Individuals carrying the MC4R-mutation are almost "resistant" to any diet and physical activity. Is it not a "faulty" preventive strategy to give advice to these individuals that "five a day" or "a low-fat diet" will be effective? Would it not be the "better" (preventive) strategy to give societal support by respecting them as they are? Of course, there are many more polymorphisms involved in obesity, and there are several polymorphisms that play an important role in the effectiveness of diet and sports. There are even polymorphisms that increase the risk of dying after physical activity. It should be kept in mind that one must be careful about the message "prevention and health promotion is good for everybody," for example, in terms of a specific diet or physical activity. In this context, the "right not to know" and the "right to know" deserve unbiased attention and must be mutually assured.[39] This has so far not been considered in most European discussions about the regulation of genetic tests. Besides questions of reimbursement and access to genetic tests or genome-based information, restrictions in the provision of genetic tests such as a physicians' proviso, which has already been considered in some countries like Germany, seem to be sheer naïveté in the era of e-health, globalization and integrated health services. Instead of proclaiming (ineffective) restrictions, would it not be much more effective and efficient to promote health literacy in order to protect the consumer?[40] And from an ethical point of view, would it not perhaps be more appropriate to use the model of "informed contract,"[41] which is based on the idea of "benefit sharing" between the consumer and the provider, instead of continuing to use the model of "informed consent" and "informed choice" in the doctor-patient relationship?

New genome-based information and technologies will force health communities to enhance surveillance (e.g. biobanks) and epidemiologic capacity for collecting and analyzing information stemming from community-based assessments of genomic variation,[42] thereby providing evidence about the burdens of various diseases. As with other fast-paced scientific and technological advances, the intersection between genomics and public policy will continue to require close monitoring, using public health methods like health technology assessment (HTA),[43] health needs assessment and health impact assessment (HIA), and will continue to require timely action. Thus, we will have the chance to ensure the appropriate and responsible use of genome-based information and of these new technologies.[44]

In summary, the following eight public health genomics issues and priorities can be identified:

1. *risk stratification and risk communication*

- earlier and higher precision of risk strata (distinction and identification of high, moderate and low risk groups; "genome-based standardisation" in addition to age and sex standardisation of diseases)
- the role of genetic determinants not only within a group of other health determinants (e.g., social, behavioural, environmental, biological) but also as a modifier and triggering factor
- the concept of a genetic variant in different individuals as a risk factor and a protective factor at the same time
- the shift from disease orientation to risk orientation and even to "disease cluster or health outcome cluster" orientation
- genetic determinants as "necessary but not sufficient" determinants in the development of complex diseases and health problems

2. *prevention*

- evidence-based primary, secondary and tertiary prevention by integrating genome-based knowledge (for example, osteoporosis or infectious diseases)
- stratified prevention by identifying high, moderate and low risk groups instead of "one prevention strategy for all" ("prevention paradox": low genetic penetrance and high frequency of genetic susceptibilities as the specific business of public health genomics)
- earlier prevention based on genome-based knowledge ("individual profiling") (for example, newborn screening as a biobank of health information starting at the beginning of life)
- minimising "faulty prevention" (for example, vaccination or sports and sudden death)
- anti-discrimination by higher target-orientation based on genomics (for example, obesity, drug and alcohol addiction)

3. *surveillance*

- recognition of well established newborn screening as an already existing nationwide DNA biobank (in public or in private hands)
- integration of genome-based biobanks into the many already existing population-based surveillance systems (e.g. cancer registries, surveillance of infectious diseases, EUROCAT (European Surveillance of Congenital Anomalies), ALSPAC (Avon Longitudinal Study of Parents and Children) or even health observatories) and in future surveillance systems covering health problems over whole lifespans
- surveillance of samples (DNA and other biomarkers as well as tissue) and data at the same time
- linkage of records (e.g. perinatal quality assurance programmes, hospital discharge data) and data from registries (e.g. cancer registries) with data from (genome-based) samples in addition to mega-(population-based) biobanks

4. *(genetic) inequalities in health*

- inequalities in genetic variants between individuals ("we are all at risk for something?")
- inequalities in genetic variants within and between populations (stratified screening programmes instead of one population screening for all) (e.g., specific migrants' needs)
- inequalities in access to genetic services
- inequalities in access to genome-based knowledge and technologies at the global level
- inequalities in the reimbursement of "genetic tests" as well as in genome-based biomarkers and technologies
- inequalities in health literacy regarding the complexity of genome-based knowledge ("widening the gap")

5. *regulations, good governance and ethics*

- balance between individual responsibility and social welfare (for example, reimbursement by sickness funds)
- strong protection of individual interests while enabling society to benefit from genome-based advances (for example, employment and occupational health)
- balance between the "right to know" and the "right not to know" (for example, European national laws on genetic diagnostics as examples of ignoring and overriding these balances)
- rethinking the principles of social justice, solidarity and subsidiarity at the individual and institutional level
- PHELSI (Public Health ELSI) (e.g. analysing and assuring demands versus needs, norms, values, preferences, health literacy)

6. *consumer protection*

- marketing and sale of genetic tests, genetic services and genome-based technologies
- food directives on the safety and quality of genetically modified food

- labelling of genetically modified food

7. *health protection*

- nutrigenomics
- recombinant vaccines (safer, cheaper and target-oriented for subpopulations)
- bioremediation (water and soil pollution)
- toxicogenomics
- envirogenomics

8. *stakeholders' responsibilities*

- rethinking of stakeholders' responsibilities in integrated healthcare (defining the responsibilities of actors and institutions in the clinical setting, primary healthcare and the public health sector)
- training of all professionals in the healthcare system in genome-based knowledge and technologies
- counselling and empowerment of the public

4.   THE EUROPEAN PERSPECTIVE

Considering genetic determinants as a factor that contributes to health and, as such, as a component for public health, is a necessary step to enable good health for all. Thus, genetic determinants must play an eminent role in any new European Union (EU) health strategy. To create sound genome-based policies and programmes, public health should get involved and, moreover, take the lead by applying the three core functions of public health (assessment, policy development and assurance) to the provision of not only genetic healthcare services but also all healthcare services.

The European Commission has, in its report on "Life Sciences and Biotechnology" (COM(2004) 250, April 7th 2004), committed itself to achieve high quality in genetic testing and to increase "co-operation and exchange of information in order to enhance coherence and disseminate best practice." Furthermore, in the 2005 work plan for "community action in the field of public health," the European Commission called for the application of a "networking exercise … to lead to an inventory report on genetic determinants relevant to public health. This network will identify public health issues linked to current national practices in applying genetic testing and on that basis will contribute to developing best practice in applying genetic testing."

Thus, in the beginning of 2006, the Public Health Genomics European Network (PHGEN), which is coordinated by the Institute of Public Health North Rhine-Westaphalia (lögd) in Bielefeld, Germany, was funded by the European Union (www.phgen.nrw.de) (EU Project No 2005313). Associated partners are the Public Health Genetics Unit (PHGU) in Cambridge, UK, as well as the German Center for Public Health Genomics (DZPHG) at the University of Applied Sciences in Bielefeld, Germany. PHGEN involves experts as collaborating partners from the fields of public health and epidemiology, human genetics and molecular biology, social sciences, (public health) ethics, medicine, economics, political sciences and (European) law. From all EU member states, applicant countries and EFTA-EEA (European Free Trade Association – European Economic Area) countries, at least there is a representative from public health and genetics as well as from a relevant competent authority. Furthermore, representatives of other European networks (e.g. EuroGentest, Orphanet, EUnetHTA or NuGO) as well as representatives of relevant initiatives and institutions on the European and international level such as the World Health Organization (WHO), the World Trade Organization (WTO), the Organisation for Economic Co-operation and Development (OECD), Scientific and Technological Options Assessment (STOA), the Agence d'évaluation des technologies et des modes d'intervention en santé (AETMIS), Centers for Disease Control and Prevention (CDC) Office of Genomics and Disease Prevention, the Genome-based Research And Population Health International Network (GRaPHInt), HumGen, TOGEN or UK DNA Banking Network are involved to ensure complementarity and to promote synergy.[45]

The aims of PHGEN are:

- to conduct a networking exercise on PHG covering all EU Member States, Applicant Countries, and EFTA-EEA Countries;
- to identify and list key experts and institutions relevant to PHG in these countries;
- to provide an inventory of genetic determinants relevant to public health;
- to provide an inventory of PHG issues and priorities in Europe;
- to identify legal diversity and barriers in a cross-border market;
- to analyse the relevance of EU treaties for PHG;
- to contribute to co-operation and information exchange in order to enhance coherence and disseminate best practice in Europe; and
- to promote and stimulate countries' efforts in this emerging field by developing PHGEN and by supporting effective networking in order to reach sustainability (e.g., implementation of National Task Forces on PHG).

In the long run, PHGEN will serve the European Commission as an "early detection unit" for horizon scanning, fact finding, and monitoring of the integration of genome-based knowledge and technologies into public health.

According to the already well-established public health trias, PHGEN's tasks include:

1. *Assessment* (the systematic collection, assembly and analysis of genome-based information and technologies relevant to public health):

   - analysis of PHG concepts (e.g. definitions of PHG, genetic determinants, genome-based knowledge, risk stratification);
   - identification of PHG issues and priorities;

- identification and "best practice" of PH methods relevant to PHG (e.g. HNA, HTA, HIA/PIA); and
- identification of networks and institutions relevant to PHG on the national, European and global level.

2. *Policy Development* (the development of European standards and guidelines which promote the responsible and effective use of genome-based information and technologies in European health systems):

  - analysis of legal diversity (e.g. conflicting laws) and barriers in a cross-border market;
  - analysis of EU treaties for PHG;
  - analysis of European minimal standards, guidelines and laws;
  - analysis of economic implications and PHELSI; and
  - development of policies on education, information and empowerment.

3. *Assurance* (the appropriate use of genome-based information and technologies in European health services):

  - critical proof of the need for enforcement of new laws and/or regulations (e.g., in most European countries there is already overregulation);
  - assurance of stakeholders' responsibilities in the application of genome-based information and technologies;
  - assurance of a competent workforce; and
  - evaluation of health services (e.g. health promotion, disease prevention, therapy, rehabilitation).

With this network, across all of Europe, there will be the opportunity for scientific advances to be translated into evidence-based policies and interventions that improve population health in a timely, effective, efficient and socially acceptable manner.

5.   CONCLUSION

What consequences can be drawn from genome-based knowledge and technologies and how can they be responsibly and promptly translated into policies and practice for the benefit of population health?[46] The necessity of assessing health services as well as of analysing complex new problems such as "genetic inequalities" in health or the role of biobanks in surveillance systems support the idea that public health should get involved and, moreover, take a leading role. Likely benefits as well as potential risks of the integration of genomics into public health interventions (assessment) should be identified. Systematically, the framework or corridors for effective, efficient and socially acceptable policies should be described (policy development) and steps and ways should be proposed to assure these policies in public health practise (assurance).[47] This will be a doable project,[48] but will require regional as well as European and global coordination.[49] There is an ethical obligation to prepare society to meet this challenge and to take up the opportunities provided by science in a medically useful, effective, efficient, socially desirable and ethically justifiable manner. Health literacy, health communication and empowerment in managing risks are key to opening the doors to a truly beneficial public health genomics. All in all, this can be facilitated by implementing ethical benchmarks such as respect for autonomy and social justice in the context of policy development.

By promoting communication about genomics in this way, not only within the public health scientific community but also among other professional groups, public health agencies and the public, perhaps there will be a return on public investment in human genome research. There are already many more opportunities than risks in providing better health for the population.[50]

Indeed, there is still discussion about stigmatization and discrimination due to genome-based information, not only among the public but also in the scientific community. Nevertheless, whoever continues separating genome-based knowledge from other medical information by defining genome-based knowledge as exceptional, whoever continues promoting the obsolete idea of genetic determinism, and whoever continues claiming the "genetization," "molecularization" and "medicalization" of society, has not seriously tried to keep up with genomic research. Explicitly, it must be emphasized, this accusation does not necessarily imply that public health professionals do not have the obligation to consider genome-based information as a highly sensitive factor in medical information. Furthermore, it is not a question of whether the combination of public health and genomics is dangerous.[51] The key question is whether, rather, harm is done to people by omitting to integrate genome-based knowledge and technologies into public health interventions and thus withholding the potential of stratified evidence-based prevention and policymaking. The public health community will lose credibility if, on the one hand, public health promotes health literacy in a society that is pluralistic and democratic in its values and enables and empowers individuals for decision-making while, on the other hand, it ignores and withholds genome-based knowledge and technologies, and therefore does not provide evidence-based public health interventions. In terms of the individual's "right to know" and in terms of best practice in public health, is this not a new form of discrimination?

The next decade will provide a window of opportunity to establish infrastructures, across Europe and globally, that will enable scientific advances to be effectively and efficiently translated into evidence-based policies and interventions that improve population health. Policymakers now have the opportunity to protect consumers, to monitor the implications of genomics for health services, and to assure that genomic advances will be taped to prevent disease and improve health. We now have the chance to prepare public health professionals, the public and policymakers for the changes to come. The above examples demonstrate approaches for the national, European and international institutionalisation of public health genetics that serve the aim of championing these challenges.

This chapter is also a result of the work of the Public Health Genomics European Network (PHGEN), which is funded in the

Public Health Programme of the European Commission (Project Number 2005313).

REFERENCES

1. L. Peltonen & V. A. McKusick, *Genomics and Medicine: Dissecting Human Disease in the Postgenomic Era*, 291 SCIENCE 1224 (2001).
2. F. S. Collins et al., *New Goals for the U.S. Genome Project: 1998-2003*, 282 SCIENCE 682 (1998); B. Childs & D. Valle, *Genetics, Biology and Disease*, 1 ANN. REV. GEN. HUM. GENET. 1 (2000); F. S. Collins & V. A. McKusick, *Implications of the Human Genome Project for Medical Science*, 285 J. A. M. A. 540 (2001); W. Burke, *Genomics as a Probe for Disease Biology*, 349 N. ENGL. J. MED. 969 (2003); D. L. Ellsworth et al., *Emerging Genomic Technologies and Analytic Methods for Population – and Clinic-Based Research*, *in* 17 Human Genome Epidemiology. A Scientific Foundation for Using Genetic Information to Improve Health and Prevent Disease (M. J. Khoury et al. eds., 2004).
3. World Health Organisation, Genomics and the World Health. Report of the Advisory Committee on Health Research (2002); H. Thorsteinsdottier et al., *Genomics Knowledge*, *in* 137 Global Public Goods for Health. Health Economic and Public Health Perspectives (R. D. Smith et al. eds., 2003).
4. P. W. Yoon, Public *Health Impact of Genetic Tests at the End of the 20th Century*, 3 GEN. IN MED. 405 (2001); A. Brand & H. Brand, *Public Health Genetics – Challenging "Public Health at the Crossroads"*, 2 ITAL. J. PUBL. HEALTH 59 (2005).
5. A. Brand et al., Gesundheitssicherung im Zeitalter der Genomforschung – Diskussion, Aktivitäten und Institutionalisierung von Public Health Genetics in Deutschland, Gutachten zur Bio- und Gentechnologie (2004); M. J. Khoury et al., *Genetics in Public Health: A Framework for the Ingegration of Human Genetics into Public Health Practice*, *in* 3 Genetics and Public Health in the 21st Century (M. J. Khoury et al. eds., 2000); R. Zimmern & C. Cook, Genetics and Health. Policy Issues for Genetic Science and their Implications for Health and Health Services (2000); G. S. Omenn, *Public Health Genetics: An Emerging Interdisciplinary Field for the Post-Genomic Era*, 21 ANN. REV. PUBLIC HEALTH 1 (2000).
6. G. Walt, Health Policy: An Introduction to Process and Power (1994).
7. M. J. Khoury, *Relationship Between Medical Genetics and Public Health: Changing the Paradigm of Disease Prevention and the Definition of a Genetic Disease*, 17 AM. J. MED. GEN. 289 (1997).
8. P. A. Baird, *Identification of Genetic Susceptibility to Common Diseases: The Case for Regulation*, 45 PERSPECTIVES BIOL. MED. 516 (2000).
9. E. Lai et al., *Medical Applications of Haplotype-Based SNP Maps: Learning to Walk Before we Run*, 32 NATURE GEN. 353 (2002); R. A. Gibbs et al., *The International HapMap Project*, 426 NATURE 789 (2003).
10. A. E. Guttmacher & F.S. Collins, *Genomic Medicine – A Primer*, 347 NEW ENGL. J. MED. 1512 (2002).

11. A. Antonovsky, Unraveling the Mystery of Health. How People Manage Stress and Stay Well (1987).
12. C. F. Sing et al., Genes, *Environment, and Cardiovascular Disease*, 23 ART. THROMB. VASC. BIOL. 1950 (2003).
13. J. S. Dorman & D. R. Mattison, Epidemiology, Molecular Biology and Public Health, in 103 Genetics and Public Health in the 21st Century. Using Genetic Information to Improve Health and Prevent Disease (M. J. Khoury et al. eds., 2000); J. Little, *Reporting and Review of the Human Genome Epidemiology Studies, in* 168 Genetics and Public Health in the 21st Century. Using Genetic Information to Improve Health and Prevent Disease (M. J. Khoury et al. eds., 2000).
14. M. J. Khoury, *From Genes to Public Health: The Applications of Genetic Technology in Disease Prevention*, 86 AM. J. PUBLIC HEALTH 1717 (1996); A. Brand, *Prädiktive Gentests – Paradigmenwechsel für Prävention und Gesundheitsversorgung?*, 64 GESUNDHEITSWESEN 224 (2002); M. E. French & J. B. Moore, Harnessing Genetics to Prevent Disease and Promote Health. Partnership for Prevention (2003).
15. Walt, *supra* note 6.
16. Brand et al., *supra* note 5; Zimmern & Cook, *supra* note 5; Michigan Center for Genomics & Public Health, Ethical, Legal and Social Issues in Public Health Genetics (PHELSI) (2004), http://www.sph.umich.edu/genomics/media/ subpage_autogen/PHELSI.pdf.
17. Omenn, *supra* note 5.
18. L. M. Beskow et al., *The Integration of Genomics into Public Health Research, Policy and Practice in the United States*, 4 COMMUNITY GENET. 2 (2001); A. Brand, *Public Health and Genetics – Dangerous Combination? View-Point Section*, 15 EUR. J. PUBL. HEALTH 113 (2005).
19. Little, *supra* note 13.
20. Brand et al., *supra* note 5; R. Zimmern, *Genetics, in* 131 Perspectives in Public Health (S. Griffiths & D. J. Hunter eds., 1999); T. Murray, *Genetic Exceptionalism and "Future Diaries": Is Genetic Information Different from Other Medical Information?, in* 60 Genetic Secrets: Protecting Privacy and Confidentiality in the Genetic Era (M. Rothstein ed., 1997).
21. M. J. Khoury et al., *Human Genome Epidemiology: Scope and Strategies, in* 3 Human Genome Epidemiology. A Scientific Foundation for Using Genetic Information to Improve Health and Prevent Disease (M. J. Khoury et al. eds., 2004).
22. P. A. Peyser & T. L. Burns, *Approaches to Quantify the Genetic Component of and Identify Genes for Complex Traits, in* 38 Human Genome Epidemiology. A Scientific Foundation for Using Genetic Information to Improve Health and Prevent Disease (M. J. Khoury et al. eds., 2004).
23. Brand et al., *supra* note 5; Khoury et al., *supra* note 21.
24. Beskow et al., *supra* note 18.
25. C. J. Frankish et al., *Challenges of Citizen Participation in Regional Health Authorities*, 54 SOC. SCIENCE & MED. 1471 (2002).
26. O. O'Neill, Autonomy and Trust in Bioethics (2002); M. Geier & P. Schröder, *The Concept of Human Dignity in Biomedical Law, in* 146 Frontiers of

European Health Law: A Multidisciplinary Approach (J. Sándor & A. P. den Exter eds., 2003).

27. Brand et al., *supra* note 5; Beskow et al., *supra* note 18; A. I. Tauber, *Sick Autonomy*, 46 PERSPECTIVES BIOL. MED. 484 (2003); UNESCO, *International Declaration on Human Genetic Data* (2003).

28. W. Burke et al., *Genetic Test Evaluation. Information Needs of Clinicians, Policy Makers and the Public*, 156 AM. J. EPIDEMIOL. 311 (2002).

29. Brand, *supra* note 14; J. M. G. Wilson & G. Jungner, Principles and Practice of Screening for Disease (1968); W. Burke et al., *Application of Population Screening Principles to Genetic Screening for Adult-Onset Conditions*, 5 GENETIC TESTING 201 (2001); M. J. Khoury et al., *Population Screening in the Age of Genomic Medicine*, 348 NEW ENGL. J. MED. 50 (2003); European Commission, 25 Recommendations on the Ethical, Legal and Social Implications of Genetic Testing (2004).

30. Brand, *supra* note 18.

31. P. Dabrock, *Capability-Approach und Decent Minimum. Befähigungs-gerechtigkeit als Kriterium möglicher Priorisierung im Gesundheitswesen*, 46 ZEITSCHRIFT FÜR EVANGELISCHE ETHIK 202 (2001).

32. R. G. Wilkinson, Unhealthy Societies. The Afflictions of Inequality (1996).

33. Murray, *supra* note 20.

34. L. P. ten Kate, *Editorial*, 1 COMMUNITY GENET. 1 (1998); L. P. ten Kate, *Community Genetics in The Netherlands*, *in* 291 Genetics and Public Health in the 21st Century. Using Genetic Information to Improve Health and Prevent Disease (M. J. Khoury et al. eds., 2000).

35. R. J. Donaldson & L. J. Donaldson, Essential Public Health Medicine (1998).

36. IoM (Institute of Medicine), The Future of Public Health. Division of Health Care Services. Committee for the Study of the Future of Public Health 7-8 (1988).

37. Brand, *supra* note 18.

38. G. Rose, The Strategy of Preventive Medicine (1992).

39. Brand et al., *supra* note 5; P. Schröder, Gendiagnostische Gerechtigkeit. Eine ethische Studie über die Herausforderungen postnataler genetischer Prädiktion (2004).

40. *Id.*

41. H. M. Sass, A *'Contract Model' for Genetic Research and Health Care for Individuals and Families*, 11 EUBIOS J. ASIAN INTERNAT. BIOETHICS 130 (2001).

42. G. J. Annas, R*ules for Research on Human Genetic Variation – Lessons from Iceland*, 342 NEW ENGL. J. MED. 1830 (2000).

43. H. D. Banta & B. R. Luce, Health Care Technology and its Assessment. An International Perspective (1993); R. J. Pollitt et al., Neonatal Screening for Inborn Errors of Metabolism: Cost, Yield and Outcome (1997); A. Brand, *Health Technology Assessment als Basis einer Prioritätensetzung*, *in* 158 Prioritätensetzung im deutschen Gesundheitswesen (B. Fozouni & B. Güntert eds., 2002); C. Moldrup, *Medical Technology Assessment of the Ethical, Social, and Legal Implications of Pharmacogenomics. A research Proposal for an Internet Citizen Jury*, 18 INT. J. TECHNOL. ASSESS. HEALTH CARE 728 (2002); M. Perleth, Evidenzbasierte Entscheidungsunterstützung im Gesundheitswesen. Konzepte und Methoden der systematischen Bewertung medizinischer

Technologien (Health Technology Assessment) in Deutschland (2003); Agence d'évaluation des technologies et des modes d'intervention en santé, Health Technology Assessment in Genetics and Policy-making in Canada : Towards a Sustainable Development. Report from a Symposium held September 11th and 12th, 2003 in Montreal (2003).

44. S. Shani et al., *Setting Priorities for the Adoption of Health Technologies on a National Level – The Israeli Experience*, 54 HEALTH POLICY 169 (2000).

45. X. Bosch, *Group Ponders Genomics and Public Health*, 295 J. A. M. A. 1762 (2006); L. Antonopoulou & P. van Meurs, *The Precautionary Principle within European Union Public Health Policy. The Implementation of the Principle under Conditions of Supranationality and Citizenship*, 66 HEALTH POLICY 179 (2003).

46. W. Burke et al., *The Path from Genome-Based Research to Population Health: Development of an International Public Health Genomics Network*, 8 GENETICS IN MEDICINE 451 (2006).

47. A. Brand et al., *Getting Ready for the Future: Integration of Genomics into Public Health Research, Policy and Practice in Europe and Globally*, 9 COMMUNITY GENET. 67 (2006).

48. G. D. Smith et al., *Genetic Epidemiology and Public Health: Hope, Hype, and Future Prospects*, 366 LANCET 1484 (2005).

49. A. S. Daar, *Top 10 Biotechnologies for Improving Health in Developing Countries*, 32 NATURE GENET. 229 (2002).

50. IoM (Institute of Medicine), Implications of Genomics for Public Health. Workshop Summary 1-67 (2005).

51. Brand, *supra* note 18.

PART III

GENOMICS AND PUBLIC HEALTH: CURRENT APPROACHES AND FUTURE PERSPECTIVES

Introduction:
The Role of International Stakeholders in Genomics and Public Health

Andrea BOGGIO

Bryant University (USA)

1. INTRODUCTION

In today's world, genomics and public health advancements are dependent upon international collaboration. Scientists collaborating at an international level carry a significant share of genomics research. Moreover, funding agencies and private funders routinely fund international projects. The World Health Organization, UNESCO, and other international institutions are key actors in the development of public health policies and strategies. Finally, patients all around the world are ultimately the recipients of the benefits of such advancements. What is the role of these different stakeholders?

Stakeholder involvement raises both theoretical and practical challenges. The interconnected nature of scientific research and the global scale of the challenges that public health concerns raise are only likely to increase the international dimension of genomics and public health activities. Moreover, differences in the cultural values, legal frameworks and political goals of the various stakeholders may have a negative impact on future development. How do we acknowledge the international dimension of genomics research and develop strategies that aim to foster international dialogue among the key actors? What groups should be represented? Who can legitimately speak for each group? How to reconcile conflicting interests? What process of involvement would ensure both efficiency and fairness?

Historically, decisions affecting the general public have been made with input from selected people—those with responsibility for the decisions or with applicable technical expertise. Increasingly, the broader public is demanding more direct involvement in decisions that will affect their lives. Other chapters in this book present a variety of points of view in order to offer insight into the role that some international stakeholders play in the complex arena of genomics and public health. In particular, these chapters describe the activities of international networks of scientists in promoting both genomics research and equitable access to its benefits, as well as the role that patients and, more generally, the public ought to play in debating the merits and justifications of conducting genomics research and its links to public health. This chapter explores the practical, ethical, and policy justifications for the role of international stakeholders in genomics and public health. It then discusses some principles and methods for a framework for stakeholder involvement in deliberations at an international level.

2.   THE BASIS OF STAKEHOLDER INVOLVEMENT IN GENOMICS AND PUBLIC HEALTH

2.1   *Practical Justifications*

Genomics is a very promising field of research when it comes to improving the health of individuals. Although genomics knowledge is also likely to benefit public health, the book of genomics' contribution to public health still needs to be written, for the most part. The scale of the scientific, economic and political challenges imposed by genomic research requires efforts involving a plurality of actors on an international level. In fact, in order to translate the potential of genomics research, time, resources and the involvement of the research community worldwide are required. Moreover, a series of practical considerations—world-wide genetic variation, the need for large-scale databases rich in data and other information from individuals with a variety of social and health backgrounds, the uncertainties surrounding the success of translating genomic research into treatments and pharmaceutical products that would enhance individuals' health, funding efforts that go beyond a single institution, and the uncertainties regarding the economic profitability of these efforts—all require coordinated and extensive efforts.

The interconnected nature of scientific research and the global scale of the challenges that public health concerns raise are only likely to increase the international dimension of genomics and public health activities. However, differences in the cultural values, legal frameworks and political goals of the various stakeholders may have a negative impact on future development. Consequently, it is important, from a practical point of view, that the international dimension of genomic research is acknowledged and that those strategies which aim to foster international dialogue among its key actors are developed.

2.2   *Ethical Justifications*

Although practical justifications based on the scale of the challenges imposed by the efforts to combine genomics and public health could alone justify involving stakeholders in discussing and deliberating genomics and public health issues, ethical reasons add support to recommending (if not requiring) stakeholder involvement at the international level. Ethical reasoning is an important step when reasoning about genomics and public health because health is a public good, and decisions concerning health can benefit and, more importantly, can harm humanity. Therefore, responsible persons ought to constantly reason about what is ethically required, permissible, or non-justifiable whenever discussing a possible course of action (policies, research programmes, or other initiatives) relating to genomics and public health. The ethical arguments that can be advanced to support this claim are based on concerns about trust, respect for the rights of the stakeholders and autonomy, with the related ideas of pluralism and deliberative democracy.

Trust is an important justification. Given that public health policies are important to various stakeholders and that public health policies often need to balance—and sometimes sacrifice—some interests against other interests, stakeholders' ability to trust policymakers to make the best possible decision is necessary for the future viability of policies affecting genomics and public health. Thus, a consequentialist perspective supports stakeholder involvement: if stakeholders trust the policymaking process, they are likely to accept its outcome even if it limits some of their interests, and genomics and public health initiatives are likely to benefit from that, as will the health of individuals. Trust is certainly fostered by stakeholder involvement, which offers opportunities for communication between policymakers and stakeholders and the possibility for the latter to express their views in the policymaking process.

Stakeholder involvement may also offer an opportunity to protect the rights of those affected by public health policies. Involving stakeholders from the community that will be affected by the public health programmes or by genomic research is certainly ethically desirable. Along these lines, the American Public Health Association's Public Health Code of Ethics, which is concerned with respecting the rights of the individuals in the community, provides that "2) Public health should achieve community health in a way that respects the rights of individuals in the community" and that "3) Public health policies, programs, and priorities should be developed and evaluated through processes that ensure an opportunity for input from community members."[1]

Autonomy is also an important ethical consideration. Granting ethical consideration to autonomy reflects the idea that each individual is the best judge of his or her own interests. As a consequence, individuals should be in the position of self-determination, of deciding their destiny for themselves. Autonomy, however, has very strong individualistic features as it has been traditionally construed as a space that ought to be protected against influences from state authority or, more generally, from collective power. Genomics and public health are areas where an individualist approach to autonomy is challenged. How to reconcile autonomy and the protection of common goods?

Expressing preferences and acting in conformity with our value system deserves ethical consideration. When dealing with public goods such as health and knowledge, collective actions must reflect individuals' desires and values. Stakeholder involvement is certainly a challenge to individualist autonomy, but it is also an opportunity for reinforcing individuals' "self ruling" prerogatives in at least two dimensions: pluralism and deliberative democracy.

First, stakeholder involvement should be seen as a means to include and mediate between diverse, culturally and value-based points of view. When dealing with research on a global scale that connects stakeholders in different parts of the world, differences in cultural values may be challenging. One strategy to cope with this challenge is to adopt a top-down approach in which principles are decided by actors—international institutions, major funding bodies, and large research institutions—in a position to impose their views on other stakeholders. A different and preferable strategy is to establish a process, that involves a plurality of stakeholders and that brings together the points of view of actors with different values, both from the centre and the periphery of genomic research. Although apparently less efficient, the latter approach seems to be ethically preferable because respect for the different value systems of the various international stakeholders is an important part of managing public goods such as health and biomedical research.

Second, stakeholder involvement allows individuals, directly or through their representatives, to participate in decisions affecting public goods so that the outcomes of deliberation reflect, to some extent, their values and desires. Indeed, democracy is often considered the best political arrangement whenever decisions affecting public goods are necessary. Deliberative democracy theories translate the too general ideal of democracy into a viable framework for democratic collective actions and decisions.[2] Emphasizing a shared process of discussion, following rules of rational argument and non-coercive exchange of views, Joshua Cohen, a prominent deliberative democracy scholar, argues that "to justify the exercise of collective political power is to proceed on the basis of a free public reasoning among equals."[3] In other words, "free public reasoning among equals" is required when health and other public goods are at stake. Within this conceptual framework, involving stakeholders in discussions and deliberations is certainly a particularly attractive strategy for discharging the duty to justify the exercise of collective political power.

The interaction between deliberative democracy and genomics and public health will be the main theme of the rest of this chapter. Different ethical considerations are in favour of the view that strategies which aim to foster international dialogue among key actors ought to be supported and implemented. Before proposing a working definition of "stakeholder" and delineating a stakeholder involvement process, this chapter will discuss the policy justifications for involving stakeholders in deliberations on genomics and public health.

2.3   Policy Justifications

In addition to practical and ethical considerations, involving stakeholders in deliberations on genomics and public health is also defensible based on two policy justifications. First, stakeholder involvement can be instrumental to a better analysis of the issues. In fact, discussing public issues helps stakeholders form opinions when they might otherwise have none or to refine and revise their views according to the views proposed by other stakeholders. The quality of decision-making should improve if it must stand up to public examination of its appropriateness and coherence. Second, stakeholder involvement usually leads participating stakeholders to commit to the deliberation process that leads to a decision and to uphold its outcomes. Thus, it increases the authoritative power of any policy outcome. As the World Health Organization recently pointed out, "The political feasibility of policy depends on: the power of the players; their position; the intensity of their commitment; and their numbers."[4] Thus, the stakeholder's "voice" of acceptance is widely perceived to add legitimacy to the decision-making process. Norman Daniels nicely summarizes the policy benefits of involving stakeholders as follows:

> "[Stakeholders] improve deliberation about relevant reasons, potentially adding to the range of considerations and the perspectives from which they are evaluated. By being involved, they take some ownership of the results, and through their potential roles as public critics or advocates, they can help explain and defend decisions they have come to take ownership for."[5]

3.   THE RELEVANT STAKEHOLDERS: A WORKING DEFINITION

Until this point, this chapter has made a case for stakeholder involvement without defining the term "stakeholder." Very simply put, the American Heritage Dictionary of the English Language defines a "stakeholder" as "One who has a share or an interest, as in an enterprise."[6] Similarly, the World Health Organization tells us that a "stakeholder" is "Any party to a transaction which has particular interests in its outcome"[7] or, alternatively, "who stands to win or lose by a line of policy."[8] Consequently, the notion of stakeholder is built around the concept of interest and policy outcome. "Interest" is defined both in terms of "right, claim, or legal share,"[9] and in terms of "participation in advantage and responsibility."[10] "Policy outcome" is synonymous with the "end result...consequence [or] effect" of a given policy.[11] In sum, this semantic excursus shows that the notion of shareholder entails the consideration of three elements: (1) a stake or interest in the outcome of a certain course of actions, (2) participation in the deliberations involving that course of actions, and (3) responsibilities towards other stakeholders.

Three implications follow from this working definition of "stakeholder." First, only actors who have an interest in the outcome of genomics research and public health policies and programmes can be listed among the relevant stakeholders. Second, if having a vested interest is the discriminating factor when it comes to including potential stakeholders among those who ought to be involved in deliberations, it also plays a critical role in the stakeholder involvement process. Stakeholders must declare what their stake is, articulate their interest in relation to the genomics and public health issue to be deliberated, and be accountable for their actions based on a declared interest to the other stakeholders. Third, rights and responsibilities follow from their inclusion among relevant stakeholders. If these individuals are entitled to be involved in deliberations which have outcomes that may affect them, they are also responsible for representing, or failing to represent, specific interests and for failing to act reasonably. Accountability is simply the reserve side of the right to be included.

4.   A LIST OF RELEVANT STAKEHOLDERS IN GENOMICS AND PUBLIC HEALTH

Based on the proposed working definition, it is possible to reason about stakeholders in genomics and public health in terms of individual and collective actors who have a stake in the outcome of genomics and public health policies and programmes. Among them, one could first name international organizations—the World Health Organization, the World Intellectual Property Organization, the World Trade Organization, UNESCO and others—that are, albeit with different powers and competencies, responsible for promoting and implementing policies in this area. Other institutional actors, at the national and international level, are also included in the category of policymakers: the US National Institutes of Health is an example of a national, governmental institution that is in the position to influence policies with international ramifications. Funding bodies should also be included in the list. The UK Medical Research Council, the Gates Foundation, and the NIH once again, have certain rights and responsibilities that follow from their involvement in funding projects on a global scale. Academics are also stakeholders because, on one hand, they contribute the most in terms of research to the field of genomics and public health and, on the other hand, they are greatly affected by policies in this area. Private companies— primarily biotech and pharmaceutical companies—have a vested, commercial interest, flowing from their investments in research and development, linked to the outcome of the research itself but also dependent upon the legal framework and the economic environment in which genomics and public health actions take place. Of course, patients are key stakeholders because, ultimately, they are the real recipients of the efforts to translate genomic knowledge into treatments and products that benefit their health. Finally, communities are important stakeholders because they are certainly affected by public health policies and programmes and also because they provide the economic (mostly through taxation) and political support for the design and implementation of genomic research and public health policies and programmes.

The interests of the listed stakeholders, however, sometimes overlap, as highlighted by Diagram 1. An example of how the line separating these categories of stakeholders is blurred is an academic who might be funded by a private company, a policymaker who could be linked to patients with specific conditions, or an institution that is inclined to favour academics doing research in its country as opposed to more qualified or better funded researchers from other countries.

DIAGRAM 1: THE STAKEHOLDERS IN PUBLIC HEALTH GENOMICS

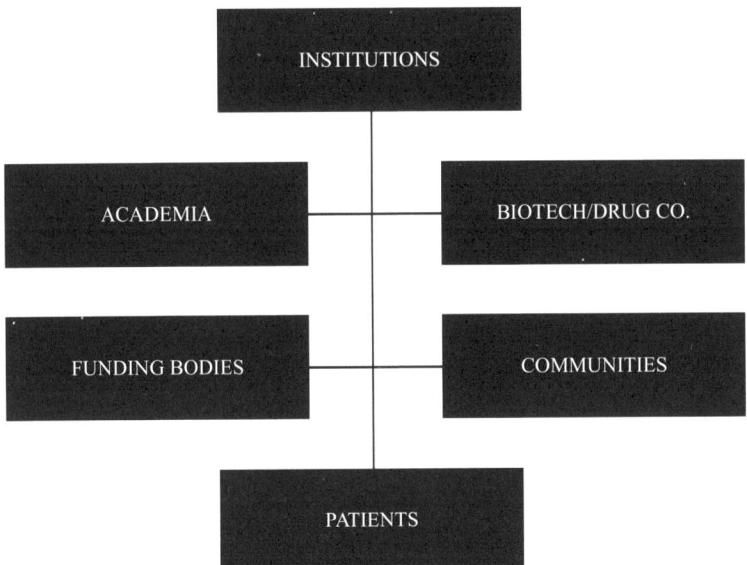

5.   THE STAKEHOLDER INVOLVEMENT PROCESS: PRINCIPLES

The final two sections of this chapter will discuss a framework for the involvement of stakeholders in deliberations concerning genomics and public health policies and programmes at an international level. The design of such a framework is challenging. On an international level, efficiency, fairness, and accountability are arguably the principles that provide the strongest foundations for the process of involving stakeholders in deliberations.

5.1   *Efficiency*

Efficiency must be construed as a function of *time* and *quality*. First, time is a critical element when goods of primary importance, such as healthcare, are at stake. Indeed, the potential benefits of genomics and the pressure imposed by public health challenges require that decisions concerning policies and programmes be reached within a reasonable delay from the time policy actions are sought. A process that requires an excessive amount of time would fail to achieve is goal, notwithstanding the merits of the policy outcomes. Second, the outcome of the deliberation process must lead to the design and implementation of policies and programmes that will effectively foster advancements in genomic research and ultimately gather information that will lead to improvement in public health measures.

5.2   *Fairness*

In this context, fairness must be intended both in its substantive and its procedural dimensions. The most important substantive trait of fairness is that stakeholder involvement improves the outcomes of the deliberation process by making them more fair. As already discussed, genomics and public health deliberations affect health outcomes, and health is a public good. Moreover, these deliberations are likely to involve certain interests. However, if more views are taken into account and more rationales are brought to the discussion in order to justify a course of action, the outcomes will likely be more fair, being grounded on a broad information base and assessed through pluralistic scrutiny.

Procedural traits of fairness encompass a variety of traits. First, fairness requires that stakeholder involvement leads to a discussion where the "majority rule does not make right." Although dissent and disagreement are likely to colour any stakeholder involvement process, the goal of such process is not to reach a consensus of more than half of the stakeholders in order for the deliberation to end, but rather through a dialogic assessment of the strength of the various interests, rationales and positions. Sometimes, an agreement in not required, and the opportunity for increasing the understanding of the various positions is a superior goal to be achieved through stakeholder involvement. Other times, formal decision-making rules apply to a specific deliberation. This is the case, for instance, in international organizations' drafting of guidelines. Specific rules regulate the process that leads to the approval of the guidelines. However, even if the process is regulated, it is important that decisions are taken not simply because "the majority says so" but rather because the propose outcome is superior from a political, technical and ethical perspective. Stakeholder involvement must foster the contribution of all points of view, and deliberation shall not ignore minority views: as long as they are rooted in reason, disagreement and dissent should be given some weight.

This leads to the second trait of procedural fairness: the non-exclusion rule. Stakeholders ought not to be excluded based on discriminatory criteria or because of socio-economic barriers (lack of funding, linguistic barriers, lack of political representation.). This rule will be relevant to the discussion of selection criteria below.

Finally, unless in its purely informal forms, stakeholder involvement must be structured around certain procedural rules that participants must agree upon. In fact, it is crucial that ground rules are set and that the selected stakeholders agree to respect them. Aiming to provide a clear framework for fair participation, the ground rules must encourage stakeholders to express reasoned perspectives on the issues being debated, balance differences in stakeholders' power and ability to influence the discussion and deliberation, indicate the scope of stakeholder involvement, and finally, indicate what decision-making process will be adopted, if decisions are required.

5.3   Accountability

As a corollary of the requirements of efficiency and fairness, involved stakeholders must be held accountable for their participation, in case they act unreasonably or non-cooperatively. Several strategies may contribute to implement accountability.[12]

The publicity requirement—information regarding the stakeholder involvement process must be publicly available, understandable by non-participating stakeholders, and aim to render the process as transparent as possible—offers the opportunity for scrutinizing the stakeholders and therefore for their accountability. By offering the possibility for the public or other stakeholders—especially those who have not been selected to participate in the process—to scrutinize the actions of the selected stakeholders, the publicity and transparency of the process lead to greater accountability. The full rationale of any recommendation or decision should also be publicly accessible.

Moreover, accountability may also be enhanced by requiring, at the outset of the process, that selected stakeholders declare and articulate the stake they have in genomics and public health, and state who they represent and their authority to do so. Such statements will set the scope of their involvement and limit the range of rationales that are permitted to serve as basis for advocating certain measures. Consequently, they will be accountable based on the declared stakes they aim to represent should they adopt an unreasonable or non-cooperative attitude.

Finally, although certainly not easy to implement, a third possible accountability strategy is the exclusion of unreasonable or non-cooperative stakeholders from deliberative processes in the future. Peer pressure is often a powerful accountability strategy. However, this strategy is easier to implement whenever stakeholders are selected by invitation, and a law or other binding regulations do not formally regulate the stakeholder involvement process.

6.   THE STAKEHOLDER INVOLVEMENT PROCESS: METHODS AND ILLUSTRATIONS

Efficiency, fairness, and accountability are principles that both inform and shape the stakeholder involvement process at an international level. The following sections describe some methods that can be used to design and implement such a process.

6.1   *Stakeholder Selection*

If efficiency is construed as a function of time and quality and is thought to be a primary concern in reasoning about genomics and public health, the stakeholder involvement process must reflect these assumptions. When it comes to the selection of stakeholders, should all stakeholders or only major stakeholders be involved? Ideally, all stakeholders would be involved. However, efficiency concerns may justify restricting involvement to stakeholders who either other, higher stakes connect to the outcome of the deliberation or to stakeholders with a wider basis of representation (for instance, a representative of a trade association rather than individual representatives from a number of companies that also belong that that trade association.)

Fairness concerns also play an important role in selecting the stakeholders to be involved. Since stakeholder involvement provides support to policy outcomes that potentially limit or sacrifice the interests of some categories of stakeholders, the opportunity to participate in public deliberation and scrutiny of the proposed policies is essential. But what does the requirement for fairness in electing participants entail? First, stakeholders should not be excluded on discriminatory criteria such as political views or membership in certain groups (religious groups, ethnic minorities). Second, stakeholders should not be excluded because of socio-economic barriers such as inability to travel due to lack of funding, inability to participate in deliberations due to linguistic barriers, or lack of political representation because a specific community is not a formal member of the UN or an identifiable nation under international law. Technological advances have made it possible to involve stakeholders in remote or poor areas of the world (for instance, by holding conferences over the internet), and therefore, alternative strategies must be explored before excluding stakeholders when socio-economic barriers prevent them from actively engaging in the discussion.

Moreover, the relevant stakeholders vary depending on the level of decision-making. If the genomics and public health deliberation involves international organizations, the representation of various geographical regions is a key factor because wider representation leads to the inclusion of stakeholders with diverse cultural and professional backgrounds, from areas with various levels of economic developmental and societal arrangements. On the other hand, if discussions concern a project affecting a specific community, representatives of community groups must take part in the process. Special consideration should be given to vulnerable groups that are potentially affected by the project. If the same project also aims to translate research in treatment, the involvement of representatives from industry and from the local ministry of health seems particularly appropriate.

6.2   Setting Ground Rules

Efficiency, fairness, and accountability also require setting ground rules that regulate the stakeholder involvement process. Efficiency requires a clear framework that limits stakeholders' actions without impeding their participation. Ground rules may include provisions regarding the timing and means of communication, the means that ensure the publicity of the process, the decision-making rules on any given issue (if required), the rules on dissent and disagreement, and finally, the rules on how to monitor and revise the policy or programme in the aftermath of the stakeholder involvement process.

Fairness requires that the ground rules favour equal participation and fair outcomes. Finally, accountability requires that participating stakeholders commit to the process, and cannot withdraw from it unless exceptional circumstances intervene.

6.3   *Publicity and Transparency*

Adequate publicity or transparency of the stakeholder involvement process is an important requirement. Deliberations concerning genomics and public health involve decisions affecting a public good, health. Therefore, securing adequate publicity or transparency of the stakeholder involvement process is certainly an important yet ambitious goal. The rationale seems even more compelling whenever international organizations or national governments are involved in the process or whenever public money is used to fund research or public health initiatives.

Ideally, the publicity requirement requires that the information be both widely accessible to the public and comprehensible by all stakeholders. If international organizations are involved, consultation organized through a form of publicly accessible hearings, including the presentation of evidence and arguments, is the primary avenue for facilitating direct participation in the process. Broadcasting or making those hearings available over the internet is an effective alternative to direct participation. Making meeting notes—including an indication of the invited parties along with their statement of interests—available for comment is also an important tool. Whenever recommendations are present, it is crucial that the rationale for the grounds of the recommendations is publicly available. Publishing in open-source journals is also an important tool that favours the involvement of stakeholders who face economic barriers to participating directly in the debate.

7.   CONCLUSIONS

Human welfare depends in part upon advancements in genomic knowledge. Consequently, genomics is strategic to public health improvement in the future. However, the challenges that genomics and public health raise are immense, and require efforts that are coordinated worldwide and that involve a variety of stakeholders. Practical, ethical and policy considerations support this claim.

Involving stakeholders in reasoning about genomics and public health is in itself challenging. Stakeholders may have interests that are economically, culturally and politically incompatible. However, the arguments in favour of opening debates and deliberation concerning an important public good such as health are compelling. The challenge is to design a framework that supports this vision. On an international level, efficiency, fairness and accountability are principles that provide the strongest foundations for the process involving stakeholders in deliberations. Those principles also provide practical guidance on how to implement stakeholder involvement with regard to the selection of stakeholders, setting ground rules that govern such a process, and its publicity and transparency. Involving all relevant actors in debates over genomics and public health is a book that, for the most part, has yet to be written. However, although the involvement of all relevant stakeholders is proving to be challenging, the lack of it would be a loss for humanity.

REFERENCES

1.   American Public Health Association, Public Health Code of Ethics (2002), http://www.apha.org/codeofethics/ethics.htm .
2.   M. Cooke, *Five Arguments for Deliberative Democracy,* 48 POLITICAL STUDIES 947 (2000).
3.   J. Cohen, *Procedure and Substance in Deliberative Democracy, in* Deliberative Democracy: Essays on Reason and Politics 412 (J. Bohman & W. Rehg eds., 1997).
4.   World Health Organization, The World Health Report 2000. Health Systems: Improving Performance 134 (2000).

5. N. Daniels, How to Achieve Fair Distribution of ARTs in "3 by 5": Fair Process and Legitimacy in Patient Selection (Prepared for WHO consultation) 21 (2004), http://www.who.int/ethics/en/background-daniels.pdf.
6. American Heritage Dictionary of the English Language (4th ed. 2000).
7. World Health Organization, A Quick Reference Compendium of Selected Key Terms Used in The World Health Report 2000 (2000), http://www.who.int/health-systems-performance/docs/whr_2000_glossary.doc.
8. World Health Organization, *supra* note 4, at 134.
9. American Heritage Dictionary of the English Language, *supra* note 6.
10. Merriam-Webster Online Dictionary, http://www.merriam-webster.com/cgi-bin/dictionary?book=Dictionary&va=interest&x=0&y=0.
11. American Heritage Dictionary of the English Language, *supra* note 6.
12. R. Loewenson, Public Participation in Health Systems (Equinet Policy Series no. 6) (2000).

From Genomic Research to Public Health Practice: International Policy Implications[1]

Stuart HOGARTH

Cambridge University, U.K.

Concerns about the quality of genetic testing have been considered in a range of policy forums, and successive policy reports have made recommendations about how to develop the regulatory regime for genetic testing. This chapter focuses on one crucial aspect of the transition of tests from the research setting to routine clinical use: the regulatory framework for the evaluation of novel tests. Over the past 15 years, as experts began to predict that genetic testing would play a greater role in disease prevention, management and treatment, there have been growing concerns that some genetic tests are entering clinical practice prematurely. As one senior diagnostics industry figure has described it:

> "[There has been] a noticeable lack of consensus within the genetics community about exactly when a test for a new marker was sufficiently validated for it to enter into clinical service. Some labs rushed to provide testing after the first publication, while others waited until the result had been replicated in multiple studies or multiple ethnic groups."[2]

The problem Emily Winn-Deen poses is not restricted to the USA; it transcends national boundaries. But does it have an international solution? The framework of regulation which governs genetic testing varies across countries/regions (and sometimes within countries where the central or federal government does not have full control over activities at the state or province level). This chapter will examine these differences and explore the opportunities for international policymaking.

Why is there a problem? Put simply, the progress of biomedical science has outstripped the rather limited mechanisms we have for evaluating diagnostic tests.[3] Traditionally, new diagnostic tests have entered clinical practice in a gradual process largely dependent on informal mechanisms of professional evaluation. By the 1990s, the inadequacies of this system were under challenge from the speed of innovation in molecular diagnostics. Whilst statutory regulation of diagnostic tests has gradually increased, and evaluation through health technology assessment is also growing, these mechanisms are failing to keep up with the proliferation of new testing technologies and new biomarkers. This problem will become greater as genetic tests move from the area of single-gene disorders and into the arena of common complex diseases. This is a fairly immediate prospect—a range of companies are preparing to come to market in 2006 with tests in areas such as heart disease, cancer and autism (see table one).

TABLE 1

COMMON COMPLEX DISEASE TESTS – AN EMERGING MARKET	
COMPANY	DISEASE
DeCode/Illumina	Heart disease
Celera	Heart disease
Jurilab/Nanogen	Heart disease
Integragen	Autism
Multiple companies	CYP450 pharmacogenetic tests

Tests for common complex disorders will challenge the existing governance framework for genetic tests:

- They will not be delivered through genetics clinics
- They will not be provided with detailed pre- and post-test counselling

• Existing regulations governing the delivery of genetic tests may not apply

There will be a new generation of genetic susceptibility tests developed in the near future. What will patients need to benefit from them? What will they have to guide them? The technologies may be new, but the answers are old ones.

> "As a patient ... I'm only going to be interested in one thing: is it a true positive or negative or is it a false positive or negative? And if it's a true positive or negative what are we going to do with that information to make me either stay well or be better."[4]

Furthermore, one could argue that we have been thinking about future patients for quite a long time. There has been a prolonged policy debate about how best to ensure the safe and appropriate use of clinical genetic tests; a number of committees and task forces have reviewed the oversight of genetic testing, and their reports have come to similar conclusions: *genetic tests should not enter routine clinical practice without thorough independent evaluation.*

MAJOR POLICY REPORTS[5]

US

1975 – *Genetics screening programmes, principles and research (National Academy of Sciences)*

1994 – *Assessing genetic risks* (Institute of Medicine)

1999 – *Promoting safe and effective genetic testing in the United States* (Task Force on Genetic Testing)

2000 – *Enhancing the oversight of genetic tests: recommendations of the Secretary's Advisory Committee on Genetic Testing* (SACGT)

UK

1994 – *Genetic screening – ethical issues* (Nuffield Council on Bioethics)

2000 – *Genetics and health – policy issues for genetic science and their implications for health and health services* (Report for the Nuffield Trust)

2000 – *NHS Laboratory services for genetics* (Report for the Department of Health)

2003 – *Genes direct. Ensuring the effective oversight of genetic tests supplied directly to the public* (Human Genetics Commission)

EU

2000 – *Report of European Parliament's temporary committee on human genetics and new technologies in modern medicine*

2003 – *Towards quality assurance and harmonisation of genetic testing services in the EU* (Institute for Prospective Technological Studies)

2004 – *Ethical, legal and social aspects of genetic testing: research, development and clinical applications* (European Commission Expert Group)

Canada and Australia

2002 – *ALRC 96 essentially yours: the protection of human genetic information in Australia* (Australia Law Reform Commission and Australian Health Ethics Committee)

2001 – *Genetic services in Ontario: mapping the future* (Provincial Advisory Committee on New Predictive Technologies)

International

2001 – *Genetic testing: policy issues for the new millennium* (OECD)

2005 – Quality assurance and proficiency testing for molecular genetic testing: summary report of a survey of 18 OECD member countries (OECD)

Furthermore, it has become a well-established view that full evaluation requires evidence on four criteria set out in the ACCE framework:

Analytic validity – accuracy of the test in identifying the biomarker

Clinical validity – relationship between the biomarker and clinical status

Clinical utility – likelihood that the test will lead to an improved outcome

Ethical, legal and social implications

However, putting this consensus view into practice has proved difficult for a number of reasons:

- the lack of standards for test evaluation;
- the lack of platforms and processes for test evaluation;
- lack of political will to plug gaps in regulation;
- the need to strike a balance between ensuring proper evaluation and encouraging innovation and access; and,
- the lack of clarity on the respective roles of different gatekeepers.

On this last point, we can think of the pathway from bench to bedside as controlled by a series of gatekeepers, creating a regulatory regime with multiple points of control. These can be seen as operating at three levels: statutory controls, resource allocation and clinical governance.[6] So the use of a genetic test might be regulated at:

- the first level, by standards set by a statutory licensing body such as the Food and Drug Administration (FDA) in the USA;
- the second level, by the requirements established by a purchaser, commissioner or reimburser of services, such as the UK's National Health Service (NHS);
- the third level, by the rules and guidelines set by professional bodies, healthcare organisations and other groups, which set standards in the practice of medicine.

Working out the respective role of each level of regulation is an urgent and important policy challenge for both national and international policymakers.

There has been a lot of policy work which points to gaps in the existing regulatory framework and makes recommendations about how to address these gaps. Some of these recommendations have been implemented but many have not. We shall now outline the existing regulatory framework, looking at three crucial questions:

1. Does independent pre-market evaluation take place? If so, how comprehensive is that evaluation?
2. What mechanisms exist for the collection and evaluation of data at the post-marketing stage?
3. What regulatory mechanisms govern the information which is provided to doctors and patients pre- and post-test?

1.   STATUTORY CONTROL

There are two overlapping statutory domains: one is the regulation of tests as medical devices, by agencies such as the FDA, and the other is the regulation of laboratory practice.

The ACCE framework does not tie in neatly with the existing authority of statutory regulators. In general, regulatory agencies do not have the statutory authority to assess the clinical utility of a test or its ethical, legal and social implications.[7] Nevertheless, the statutory framework for the regulation of medical devices can provide a robust mechanism for ensuring pre-market review of analytic validity and, to a lesser degree, clinical validity. However, even this level of review is not consistent, largely because of a number of regulatory gaps.

1.1   *Clinical Validity Evaluation*

The US and Canadian systems emphasise the importance of pre-market evaluation of clinical validity data. The European system and Australian system are focused on analytic validity, although any clinical claims must be supported with evidence—an ambiguous position which we will return to.

TABLE 2

EVALUATION OF CLINICAL VALIDITY	
USA	Yes
Canada	Yes
Europe	Only where clinical claims are made
Australia	Only where clinical claims are made

1.2 Risk Classification

The issue of risk classification is tied to pre-market evaluation because the regulatory systems for in-vitro diagnostics (IVD) tests are predicated on risk classification. Those tests which are considered higher risk, because of their clinical or public health significance, or sometimes because of their novelty, are subject to greater scrutiny. Regulatory gaps can appear when tests are deemed low-risk and therefore exempt from pre-market review.

In the United States, Canada and Australia, genetic tests are all treated as moderate—to high-risk—and so are subject to pre-market review. However, there is a major exception to this in the US. Many device manufacturers are exempt from FDA pre-market review because they sell laboratories what are termed "analyte specific reagents" (ASRs), the active ingredients of in-house tests. The FDA has classified nearly all ASRs as Class I and therefore has exempt them from pre-market review.

In Europe, genetic tests are treated as low-risk, and so are exempt from independent pre-market review. In effect, the European system does not have a workable mechanism for classifying the risk profile of novel tests; the automatic assumption is that all novel tests are low-risk.

There is a tension in the European and Australian systems between the importance of risk classification, and their emphasis on evaluation of analytic rather than clinical validity, since without a stated intended clinical use, it is not clear which risk class a test belongs in, so potentially high-risk tests may evade pre-market evaluation.

TABLE 3

RISK CLASSIFICATION OF GENETIC TESTS		
COUNTRY	RISK CATEGORIES	GENETIC TESTS
USA	I – III	Mostly II so far (but ASRs are Class I)
Canada	I – IV	III
Australia	I – IV	II or III
Europe	I – III	I

1.3   In-House Tests

Genetic testing is characterised by a high degree of dependence on tests developed in-house by laboratories. In general, the regulation of clinical laboratories is focused on quality assurance of laboratory procedures and on the analytical accuracy of laboratory testing; clinical validation of in-house tests is rarely mandatory.

In Europe and Australia, in-house tests are included in the device regulations (although there are exemptions in the European system for public health institutions). In Canada, the device regulators are considering their authority. In the US, the FDA has shifted its position several times but, as with ASRs, it seems quite likely to act in an increasing number of areas it considers high-risk: for instance, it has entered discussions with Genomic Health about their Oncotype DX test, a gene expression test for guiding breast cancer treatment.[8]

In the US, the Secretary's Advisory Committee on Genetic Testing (SACGT) recommended that the regulation of laboratory testing should be enhanced to ensure that labs provide data on the clinical validity of their tests. In recent years, the advisory committee which has oversight of the Clinical Laboratory Improvement Amendments (CLIA) regulations has been working to introduce a genetic testing specialty to develop new standards for genetic testing, including the addition of requirements for clinical validity data. However, the clinical validation aspect of this rule will not be for pre-market evaluation; instead, data on clinical validity will be examined at the time of the laboratory inspection.[9] However, given that inspections take place only every two years, then such a system cannot deliver pre-market evaluation of a test.

But, there is an alternative system of control: the State of New York has its own licensing system and requires laboratories to submit clinical validity data on new tests for pre-market approval. This system has a major impact on genetic testing because all the major US reference laboratories and many of the medium-sized ones are New York State-licensed; thus, probably at least 60% of genetic tests carried out in the US are covered by the New York State system.[10]

TABLE 4

STATUTORY PRE-MARKET REVIEW OF IN-HOUSE TESTS		
COUNTRY/STATE	CLINICAL VALIDITY	PRE-MARKET REVIEW
USA	Not yet	No
NY State	Yes	Yes
Canada	No	No
Australia	If claims are made	Yes
Europe	If claims are made	No (low-risk)

1.4   *Postmarketing Surveillance*

In the past, device regulation, like drug regulation, has tended to focus on pre-market review, but post-marketing surveillance (PMS) has taken on increasing importance in recent years. For instance, in Europe the In-vitro Diagnostics (IVD) Directive requires manufacturers to have a systematic procedure to review experience gained from their devices in the post-production phase. Guidance indicates that a PMS system should be in place to collect data on issues such as "changing performance trends [and] performance in different use populations."[11]

1.5   Statutory Control – Summary

Statutory mechanisms do provide independent evaluation and can cover both pre-market review and post-marketing surveillance. However, a number of regulatory gaps exist, some of which are fundamental—it is highly unlikely that regulators will take on the review of the ethical, legal and social implications (ELSI) of tests (although it is possible that regulators' awareness of the ELSI debate may have influenced their risk classification of genetic tests in Australia, Canada and the US). Neither are regulators likely to expand their pre-market reviews to include clinical utility (with the exception of the FDA, although here the question arises of whether their evaluation will go beyond establishing a basic plausibility for utility). By contrast, review of clinical validity is covered by statutory mechanisms but not consistently. In general, the issue is not one of authority but one of interpretation and enforcement, and there seems considerable scope for enhancement of this aspect.

Exemptions for in-house tests are another area of regulatory ambiguity where clarification is needed. Europe and Australia have sought to create a more level playing field through device regulation. The example of New York State demonstrates that laboratory regulation can be enhanced to encompass pre-market review. Furthermore, the New York State system addresses the serious issue of off-label use, because laboratories which seek to change the approved intended use must submit the new use for approval.

It is important to note one final point on the role of statutory regulation, relating to the provision of information to doctors and patients. In device regulations, test manufacturers have to have a label for their test. Just like a drug label, this summarises the test's performance characteristics and provides instructions for the user on its safe and effective application. The claims made on the label and any promotional claims should be backed up by data in the technical file which is submitted for review. An accurate and truthful label is one of the main things which statutory regulation can achieve. However, **there is no equivalent of a label for in-house tests**. Even in the Australian and European systems where in-house tests now fall under the device regulations, this issue has not been addressed.

2.   REIMBURSEMENT

Some of the regulatory gaps which exist at the statutory level can be addressed at the level of reimbursement or resource allocation, particularly through health technology assessment (HTA) review, which provides an alternative mechanism for evidence-based evaluation. HTA reviews are broader in scope and so can encompass the evaluation of clinical utility and ELSI, as well as analytic validity and clinical validity. The debate about the regulation of genetic testing has led to two important policy initiatives in the UK and the US.

2.1   *UK Genetic Testing Network (UKGTN)*

National Health Service (NHS) services and interventions are commissioned by health authorities, and genetic tests are no exception. A system exists whereby only tests approved by the UK Genetic Testing Network (UKGTN) may be funded through such mechanisms. Since 2003, new tests, and existing tests which are deemed worthy of investigation, are evaluated via a Gene Dossier which requires evidence on a range of issues: the seriousness and prevalence of the condition being tested for, the purpose of the test, its analytic and clinical validity, its clinical utility, and ethical, legal and social considerations. Tests which meet the criteria are submitted to UKGTN for approval and are then added to the NHS Directory of Molecular Genetic Testing as Network services. Funding recommendations are made by Genetics Comissioning Advisory Group (GenCAG) to individual commissioners within the NHS. Since its introduction in 2003, around 30 genetic tests have been formally evaluated through this process. The Gene Dossier process has attracted the interest and approval of geneticists in the US and Europe.

There is no obligation for commercial labs to join UKGTN, or if they do so, to submit their tests for pre-market evaluation. One private sector lab has joined the network, but none of its tests have yet been formally evaluated by the UKGTN. The focus of the UKGTN is, at present, on tests for inherited disorders.

2.2   Evaluation of Genomic Applications in Practice and Prevention (EGAPP)

Following the recommendations of the Secretary's Advisory Committee on Genetic Testing (SACGT), the Office of Genomics and Disease Prevention (OGDP) is taking a lead role in the development of the ACCE framework for systematic test evaluation, from data gathering and analysis to dissemination. They began with a three-year project to develop and test the ACCE framework by looking at tests for five different disorders, with the goal of facilitating an appropriate transition of genetic tests from investigational settings to use in clinical and public health practice. This work is now complete, and a new project EGAPP is taking the process forward by looking at how systematic test evaluation can be used in practice. The EGAPP working group will disseminate its findings and make recommendations based on the findings of the evidence reviews they commission. Its focus is on tests which have potential for a major public health impact, so it is very interested in both pharmacogenetics and tests for common complex diseases. However, this does not represent pre-market evaluation since tests can come on to the market prior to EGAPP conducting a review.

In other countries, such as Canada, genetic test evaluation has been considered within the context of the broader HTA programmes. In France, for instance, reviews of the clinical validity and utility of genetic tests are routinely performed by the French National Authority for Health (HAS). In Australia, all new diagnostic tests must be individually assessed by the Medical Services Advisory

Council (MSAC) before they can be approved for Medicare reimbursement. However, this system does not cover the private sector.

2.3 *Reimbursement as Control – Summary*

For reimbursement to be an effective mechanism of control, there must be well-established and comprehensive systems for evaluation of new tests. Progress has been made in the UK and the US, but significant gaps remain. The broader scope of HTA review means that it has the potential to address the whole ACCE framework of evaluation. It is strong on comparative studies and can offer detailed review at the post-market stage, when further data has gathered on a test. Furthermore, the guidance it offers to doctors and patients can help to limit off-label use for which no good evidence exists. But it rarely provides pre-market review. A further weakness of HTA programmes is that they work with existing data and, where this is inadequate, there is no mechanism to generate further data. This highlights the need for an infrastructure for systematic data collection.

3. CLINICAL GOVERNANCE

Professional societies can play an important role in evaluation of new tests by developing practice guidelines which recommend when and how tests should be used. Like HTA processes, practice guidelines have the advantage that they can consider all the elements of the ACCE framework. Although an independent and generally objective form of review, practice guidelines do not provide systematic pre-market evaluation, as they do not capture all tests and often take place after a test has entered clinical practice. Furthermore, they have tended to be developed on the basis of expert opinion rather than systematic data review, although this is now changing.

Their voluntary nature means that their enforcement powers are strictly limited.

4.   CONCLUSION

None of the mechanisms and levels of regulation which we have reviewed will be sufficient in itself to satisfy the need for systematic pre-and post-market evaluation. Improving regulation will require attention to the gaps which exist at each level but also a clear model of how the different mechanisms should interact.

Consideration of this policy issue can benefit from an international discussion about the role of respective gatekeepers and what is expected of statutory review, health technology assessment and the role of professional practice guidelines.

4.1   *Relationship Between National and International Policymaking*

What we have seen thus far are the differences and similarities between national regulatory regimes. These differences arise in part from historically divergent approaches to regulation and in part from differences in national healthcare systems. Within a single region such as Europe, there are very wide differences; even within countries, the differences at the level of individual states or provinces can be significant. So what is the role of international policymakers? There are probably two main functions:

- Information gathering – provide an overview of what is happening; identify problems and opportunities
- Standard setting – use best practices as a benchmark for practice; build consensus and thus help initiate change at nation-state level

To understand the potential role of international policymaking, we should consider the current organisations, initiatives and policy-making fora which already have some stake in these issues.

4.1.1 EuroGentest

In the EU, there have been considerable efforts to harmonise non-statutory oversight of laboratory quality assurance systems. This has been developed through a number of national, regional and international schemes which culminated in the European Molecular Genetics Quality Network. Participants in such schemes include 34 European countries and labs from Australia and the USA.

These quality assurance (QA) initiatives have led to a new project – EuroGentest,[12] an ambitious attempt to move beyond the previous focus on laboratory quality assurance and to develop a series of discrete but linked programmes which deal with all aspects of quality in genetic testing services, from evaluation of the clinical validity and utility of tests to genetic counselling. EuroGentest has attracted stakeholders in the US and Australia, and members of the project are also working as part of an Organisation for Economic Co-operation and Development (OECD) expert group on international standards.

4.1.2 The Organisation for Economic Co-operation and Development (OECD)

The OECD began its work in this area with a meeting in 2000 that looked at the policy issues around genetic testing. Out of that came the formation of an expert group to work on the development of international guidelines for quality assurance in molecular genetic testing. These draft guidelines set out to ensure minimum international requirements for quality assurance systems and laboratory practices, facilitate mutual recognition of national QA frameworks, strengthen international co-operation and increase public confidence in the governance of testing. The guidelines illustrate the potential role OECD can play in developing standards internationally. These guidelines will be made public for consultation by September 2006.

The guidelines identify collecting data on clinical validity as an essential part of laboratory quality assurance. As part of this work, the

OECD organised an international gathering to look at the evaluation of clinical validity and clinical utility. Experts who are involved in the UK Genetic Testing Network, the EGAPP project in the US and the Agence d'évaluation des technologies et des modes d'intervention en santé (AETMIS) in Canada shared their experience of genetic test evaluation with stakeholders from across the OECD member countries.

The OECD is undertaking a range of policy work around innovative health technologies. As part of this programme, they recently held a two-day workshop on pharmacogenomics. Regulators interacted with industry, clinicians, academic scientists, healthcare policymakers and other stakeholders in a discussion about the policy challenges arising from this new technology. The conclusions of the meeting will be outlined in a policy report (due for publication by the end of 2006) which will be directed to government and relevant stakeholders. The OECD may initiate further policy work in this area, as part of its biotechnology programme.

4.1.3 World Health Organisation (WHO)

The WHO is participating in the OECD process and is also working on the area of medical device regulations. In 2003, they published a comprehensive overview which set out the general principles common to regulations in different countries.[13] They are now working with the Global Harmonisation Task Force (GHTF) on a range of issues.

4.1.4 Global Harmonisation Task Force (GHTF)

Somewhat equivalent to the ICH in pharmaceuticals, the GHTF brings together a number of countries committed to exploring harmonisation of medical device regulation. Although the issue of genetics is not being addressed specifically in the work of the GHTF, its activities touch on a number of issues relevant to genetic testing outlined above, such as clinical evaluation and risk classification.

In conclusion, we clearly have a great deal to learn from each other, and international policymaking forums such as the OECD provide an opportunity to work together on common policy issues. In doing so, we might want to be guided by two important priorities: transparency and stakeholder inclusion. Furthermore, we need to ensure that any moves towards harmonisation are guided by best practice to improve standards rather than seeking a lowest common denominator.

REFERENCES

1. This chapter draws on research which has been conducted as part of a project funded by the Wellcome Trust. The author is grateful to the Trust for its support.
2. E. Winn-Deen, *Fulfilling the Promise of Personalized Medicine*, IVD TECHNOLOGY, Nov. 2003, at. 16.
3. S. Ramsey et al, *Towards Evidence-Based Assessment for Coverage and Reimbursement of Laboratory-Based Diagnostic and Genetic Tests,* 12 AM. J. MANAG. CARE 197 (2006).
4. Interview with FDA Official (2004).
5. This list is not exhaustive and does not cover the many academic policy articles which have been published around this subject.
6. W. Burke & R. L. Zimmern, *Ensuring the Appropriate Use of Genetic Tests,* 5 NAT. REV. GENET. 955 (2004).
7. The one exception to the clinical utility gap is the FDA, whose statutory authority over what they term "clinical effectiveness" can extend to clinical utility.
8. GenomeWeb Daily News, *FDA Questions Whether Genomic Health Should Have Obtained Pre-Market Approval*, January 30, 2006, http://genomeweb.com. The FDA letter can be accessed at: http://www.sec.gov/Archives/edgar/data/1131324/000095013406001293/f16614exv99w1.htm.
9. Such a system has already been put in place by the College of American Pathologists (CAP), one of the third-party bodies whose inspections are recognised by the Centers for Medicare and Medicaid Services (CMS) as equivalent to CLIA.
10. Figures are based on industry estimates. Interview with in-vitro diagnostics (IVD) manufacturer (2005).
11. European Association of Notified Bodies for Medical Devices, *Post-Marketing Surveillance (PMS) Post Market/Production*, Recommendation NB-MED /2.12/Rec1, 2000, http://www.meddev.info/_documents/R2_12-1_rev11.pdf.
12. EuroGentest, http://www.eurogentest.org/cocoon/egtorg/web/index.xhtml.
13. World Health Organization, Medical Device Regulations: Global Overview and Guiding Principles (2003), http://www.who.int/medical_devices/publications/en/MD_Regulations.pdf.

GRaPH *Int*: An International Network for Public Health Genomics

Alison STEWART

Public Health Genetics Unit, Cambridge, UK

Mohamed KARMALI

Public Health Agency of Canada, Canada

Ron ZIMMERN

Public Health Genetics Unit, Cambridge, UK

1.   INTRODUCTION

It has been widely predicted that new knowledge and technologies stemming from the Human Genome Project will in time have profound implications for medicine and health care. The discipline of public health genomics aims to ensure that genomic knowledge and technologies are used responsibly to benefit population health. This chapter describes moves to establish public health genomics on an international footing, and the culmination of these efforts in the establishment of a new international network: the Genome-based Research and Population Health International Network (GRaPH *Int*).

2.   THE HISTORY OF PUBLIC HEALTH GENOMICS

The beginnings of public health genomics can be traced back to the mid-1990s, as the Human Genome Project entered an exponential phase. In the United States, a seminal paper published in 1996 by Muin Khoury in the American Journal of Public Health (*From genes to public health: applications of genetics in disease prevention*)[1] was followed in the same year by the establishment of a Task Force on Genetics and Disease Prevention by the Centers for Disease Control and Prevention (CDC) in Atlanta. In 1997, the recommendations of the Task Force's report *Translating advances in human genetics into public health action*[2] led to the establishment of the Office of Genetics and Disease Prevention (OGDP) at CDC, under Khoury's leadership.[3] The first annual conference on genetics and public health was held in Atlanta in 1997, attracting delegates from across the United States and internationally.

In the academic setting, the Universities of Washington and Michigan were quick to see the need for post-graduate training in public health genetics, and the first multi-disciplinary Masters programmes were established at those Universities during the second half of the decade.[4] These programmes were developed and supported by faculty from a variety of academic departments including: epidemiology and public health, medical ethics, law, pharmacy and social sciences, who also adopted a multi-disciplinary approach in their own research.

The second half of the 1990s also saw a growing awareness in the United Kingdom that the National Health Service could not ignore the potential influence of genetics and genomics on healthcare. Two reports to the UK Government by an expert advisory group pointed out that far-reaching changes would result from a growing understanding of the effects of normal genetic variation on susceptibility to disease, disease progression and response to treatment.[5] In 1997, mirroring developments across the Atlantic, the Public Health Genetics Unit was set up by Ron Zimmern in Cambridge.[6]

During the first years of the new millennium, the discipline of public health genetics consolidated its position in both the US and the UK. A growing body of papers in the scientific literature established a solid intellectual basis for the new discipline and the groups such as those in Atlanta, Seattle and Cambridge initiated a range of programmes, activities and collaborations aimed at establishing an understanding of genetics and its ethical and social dimensions within the profession of public health.

3.   DEFINITIONS OF PUBLIC HEALTH GENOMICS

A variety of definitions for public health genetics were developed by its early practitioners. Although these definitions differed in detail, they were broadly similar in their fundamental concepts.

The definition adopted in the UK built upon the Acheson definition of public health, defining public health genetics as:

> "The application of advances in genetics on the art and science of promoting health and preventing disease through the organised efforts of society."

This and other definitions emphasised three important points:

1.   A broad scope for the word "genetics." The term encompassed not only genetics as inheritance (implying familial associations and genetic diseases inherited in a Mendelian fashion—the province of medical genetics services) but also genetics as the basic molecular programme underlying development, normal physiology and disease; that is, the concept of genomic medicine. The need to convey this broader meaning for "genetics" has led during the last few years to a trend towards replacing it with the word "genomics." The OGDP, for example, changed its name to the Office of Genomics and Disease Prevention in 2001.
2.   The potential for using genetics/genomics in the context of disease prevention, including primary prevention (the

prevention of disease initiation), but also clinical measures to delay disease progression and reduce disability.

3. The multi-disciplinary nature of the endeavour, which operates within a social and political context and involves insights from the arts, humanities and social sciences as well as genomic and population sciences.

4. GENES AND ENVIRONMENT AS DETERMINANTS OF DISEASE

At the heart of all conceptions of public health genomics was an emphasis on the combined effects of genes and environment as determinants of health, and an insistence on moving away from the flawed and outdated "nature versus nurture" argument. "Environmental" determinants comprise a diverse array of influences including not just obvious factors such as the air we breathe or the food we eat, but also the built environment, social factors such as poverty and deprivation, and the political system and its priorities.

The realisation that all disease results from the combined effects of genes and environment has important implications for prevention. Juengst distinguished the concepts of genotypic and phenotypic prevention; that is, prevention either by altering genes, or by altering modifiable environmental factors.[7] Although genotypic prevention may be important in some contexts for Mendelian disease—for example, at-risk couples may choose to use prenatal diagnosis to avoid the birth of a child affected by a serious genetic disease—in the context of common chronic disease, only phenotypic prevention is generally feasible or ethically acceptable.

The aspiration underlying public health genomics is that it may be possible, once we understand both the genetic and the environmental factors involved in the causation of disease and how they interact, to devise effective preventive interventions targeted at individuals with specific genotypes. These preventive strategies would involve modification of one or more of the environmental determinants. This "personalised" approach can be extended to disease management as well as prevention. For example, the study of pharmacogenetics aims to elucidate the relationship between genetic factors and response to medicines, so that drugs may be targeted at those most likely to respond and least likely to suffer adverse reactions.

Although the benefits of understanding how genes and environment work together as determinants of health are profound, the complexity of the task must not be under-estimated. Individual genes act not in isolation but as components of complex control circuits that regulate gene expression patterns in different cell types, tissues and organs. These expression patterns are established and maintained by epigenetic mechanisms: chemical modifications to DNA that do not change the primary sequence of the gene but affect its transcriptional activity and thereby the function and activity of the cell. Those cellular functions and activities are actually carried out not by genes but by proteins; even though we now have the full sequence of the genome, we are far from understanding the range and properties of its protein products, which, as a result of mechanisms such as post-transcriptional and post-translational modification, outnumber their encoding genes by up to two orders of magnitude. A full understanding of biological systems must also move beyond the molecular to the cellular and systems levels, integrating the simultaneous activities of multiple gene-regulatory circuits, signal transduction systems, cell-cell interactions and long-range influences such as circulating hormones, and understanding how these systems are modulated both temporally and by myriad environmental influences.

In view of the considerable challenge that lies ahead in developing an understanding of how the genome works in health and disease, public health genomics stresses the importance of ensuring that any new tests or interventions arising from genomic research are not introduced prematurely but are thoroughly evaluated and supported by a sound evidence base.

5.   PUBLIC HEALTH GENOMICS GOES INTERNATIONAL: THE
BELLAGIO INITIATIVE

During the last few years, contacts have grown between the centres in Atlanta, Seattle and Cambridge, and groups in other countries who are engaged in programmes and activities related to the goals of public health genomics. These contacts culminated, in 2005, with the organisation of an expert multi-disciplinary workshop attended by 18 delegates from five countries (USA, UK, Canada, France and Germany). The workshop was funded by the Rockefeller Foundation and held at the Foundation's international study and conference centre in Bellagio, Italy.

The aims of the workshop were:

1.   To explore the possibility of establishing an international network to promote the goals of public health genomics;
2.   To share knowledge and resources; and
3.   To ensure equitable access to the benefits of genome-based knowledge by all, including those in developing countries.

Workshop sessions explored a range of issues as part of the process of seeking a consensus on the scope and definition of public health genomics, and on the best way of moving its agenda forward at an international level. These questions and issues included:

1.   What are the fundamental concepts of public health genomics?
2.   Can personalised medicine be reconciled with the population-level goals of public health?
3.   What are the key ethical, legal and social issues and how can they be addressed?
4.   How can different disciplines work together to achieve shared goals?
5.   What genetics competencies do health professionals need?

5.1   *The Bellagio Statement*

As a result of its deliberations, the Bellagio workshop agreed that public health genomics could be defined as:

> "The responsible and effective translation of genome-based knowledge and technologies for the benefit of population health."

The wording of the statement was chosen with great care in order to convey the precise meanings the group intended. The term "genome-based" was chosen instead of "genetic" or "genomic" to indicate that the scope of the relevant scientific research base included not only genes but also their protein products, the metabolites synthesised by those proteins, and the interactions among all the components of the biological system at the molecular, cellular and tissue/organ levels. It was important to emphasise that both knowledge *and* technologies arising from genome-based research are relevant to public health genomics, particularly in the context of biotechnologies that may bring great benefits for healthcare and disease prevention in the developing world.

The words "responsible and effective" convey the importance of an evidence-based approach. The evidence that is required includes not only scientific and clinical data on the effectiveness of new tests and interventions, but also a thorough investigation of any ethical, legal or social consequences of their use. The Bellagio workshop emphasised the need for an integrated, multidisciplinary approach that moves away from a tendency to view ELSI (ethical, legal and social issues) research as an optional extra tacked on to the end of the scientific/clinical agenda.

Although the terms "public health genetics" and "public health genomics" are now widespread, they have created some misunderstandings as a result of the very different meanings attached to public health. In some countries, the connotations of public health are negative, implying poorly-resourced healthcare programmes for deprived communities. In others, the scope of public health involvement in genetics may be largely limited to state-led initiatives such as newborn screening programmes, while in others again the term defines a broad sphere of action encompassing both the strategic planning and the organisation and delivery of health services. The Bellagio workshop participants decided that the phrase "for the benefit of population health" most clearly conveyed the broad goals of public health genomics and its involvement across the whole scientific, clinical, social and political landscape.

5.2   *Avoiding "Genetic Exceptionalism"*

The achievements of the Human Genome Project have led to understandable enthusiasm about the potential for using information about genes and DNA variation to develop new approaches to the classification, treatment and prevention of disease. Statements about the power of genetics have led to the idea that genetic information is more powerful than other types of personal medical information and merits special protection for that reason. This concept has been described as "genetic exceptionalism."[8] Concerns about the predictive power of genetic information have led in turn to anxiety about the possible misuse of this information to discriminate unfairly against individuals. Frequently, calls are made to outlaw the use of DNA test results for particular purposes, such as in the context of decisions about insurance or employment.

Although it is, of course, important to ensure that people are not subject to unfair discrimination for any reason, it is also important to encourage clarity of thinking about the actual predictive power of DNA variants, which in the context of common disease may on their own be less predictive of ill health than phenotypic biomarkers such as blood proteins and metabolites, or lifestyle factors such as smoking status or diet. It is illogical, then, to forbid the use of DNA-based risk information while placing no restrictions on other determinants of risk. Public health genomics stresses that, although genetic variants play a part in susceptibility to disease, they should not be either privileged or unreasonably demonised.

5.3   The "Enterprise" of Public Health Genomics

The Bellagio workshop developed a consensus view of the "enterprise" of public health genomics, which is represented by the shaded areas in Figure 1 (p. 270).

Several key features emerge from this representation:

1. The input to the enterprise (on the left of the diagram; not forming part of public health genomics itself) is the research base, both in genome-based science and technology and also in the population sciences, the humanities and the social sciences. This is the phase of knowledge generation. The goal of the enterprise (on the right of the diagram) is benefit for population health.

2. Information stemming from basic research is not usable on its own, but must be integrated, both within and across disciplines. Public health genomics begins with knowledge integration, defined as the process of selecting, storing, collating, analysing, integrating and disseminating information. This is the means by which information is transformed into knowledge, and is the driving force of the enterprise.

3. The integrated knowledge base for public health genomics is used to underpin four core sets of activities:
    (a) Communication and stakeholder engagement (including, for example, public dialogue and involvement, and engagement with industry)
    (b) Informing public policy (including applied legal and policy analysis, engagement in the policy-making process, seeking international comparisons and working with government)
    (c) Developing and evaluating health services (including strategic planning, manpower planning and capacity building, service review and evaluation, and development of new programmes and services)
    (d) Education and training (including programmes of genetic literacy for health professionals and generally within

society, specific training for public health genomics specialists, and development of courses and materials)

4. The mode of working of public health genomics is described by the cycle of analysis—strategy—action—evaluation, which is a widely recognised representation of public health practice.

5. Public health genomics does include a research component, shown at the bottom of the diagram. This is not basic research, but programmes of applied and translational research that both contribute directly to the goal of improving population health and also identify gaps in the knowledge base that need to be addressed by further basic research.

6. Public health genomics does not operate in a vacuum. It is embedded within a social and political context, and informed by societal priorities.

7. Double-headed arrows throughout the diagram indicate the dynamic and interactive nature of the enterprise: it generates knowledge as well as using it, and is modulated by the effects of its own outputs and activities.

This vision for public health genomics has been fleshed out in more detail in a paper by the Bellagio participants published in the journal *Genetics in Medicine*,[9] and in a full report of the workshop that may be accessed on the GRaPH *Int* website (see below).[10]

5.4   *Establishing an International Network: GRaPH* Int

Public health genomics is still in an early phase of development. It has yet to reach "critical mass" in any individual country and in many parts of the world does not yet exist. It is essential that the pioneer groups and organisations work together to share resources, provide credibility for those wishing to develop public health genomics in their own countries, and establish collaborations in key areas of work.

To this end, the Bellagio group decided to set up an international public health genomics network, to be known as the Genome-based Research and Population Health International Network, or GRaPH

Int. The term "Int" also signifies that the Network is interdisciplinary and integrated.

Echoing the Bellagio statement, a mission was agreed for the new network:

> "GRaPH Int is an international collaboration that facilitates the responsible and effective integration of genome-based knowledge and technologies into public policies, programmes and services for improving population health."

6.   GRAPH *INT* MOVES FORWARD

The Public Health Agency of Canada accepted an invitation from the Bellagio workshop to establish an administrative hub for the Network, which was officially launched by the Chief Public Health Officer of Canada, Dr. David Butler-Jones, at the 4th International DNA Sampling Conference: Genomics and Public Health, on 6 June 2006.

The initial Steering Group, defining the strategy and direction of GRaPH *Int,* comprises the members of the Bellagio workshop group. It is envisaged that this will be a transitional arrangement and that the composition of the Steering Group will broaden and assume a more truly international character during the next few years as the Network develops.

Responsibility for moving plans for the Network forward has been taken by a smaller Executive Group [(Professor Wylie Burke (Seattle), Dr. Mohamed Karmali (Guelph), Dr. Muin Khoury (Atlanta), Professor Julian Little (Ottawa) and Dr. Ron Zimmern (Cambridge)] and by a Secretariat based at the Centre de recherche en droit public at the University of Montreal. The role of the Secretariat is to support the work of the Network's Steering Group, Executive Group and Working Groups, and to develop and maintain a website for the Network.

The GraPH *Int* website aims to provide a portal to the organisations and resources that are available worldwide to support the multidisciplinary enterprise of public health genomics.[11] A navigation system for the site is being developed that uses, as its basis, the "enterprise" diagram shown in Figure 1, with links to sources of information relevant to the core activities and functions of public health genomics. A News and Views section features lively summaries of recent conferences and other events, and Q&A sessions with key individuals in public health genomics. Those interested in being informed about the work of GRaPH *Int* and developments in public health genomics can subscribe, via the website, to regular e-mail updates.

6.1   *Work Programme*

Interdisciplinary GRaPH *Int* Working Groups have been set up in three key areas. Chairs have been appointed for the groups, and discussions about initial priorities have begun. The three groups are:

1. Research (aiming to define the research needs of the enterprise and working to inform the priorities of major funders)
2. Education and training (defining competencies and identifying ways of addressing education and training needs)
3. Ethical, legal and social issues (working to achieve more effective integration within the ELSI field and between ELSI and other parts of the enterprise)

7.   CONCLUSIONS

Public health practice in the 21st century can no longer ignore the knowledge derived from genetic and molecular science. An understanding of the molecular and cellular mechanisms of disease will be as important to the public health community of the future as an understanding of the social determinants of health. Achieving benefits for population health from advances in genomic science and technology will depend on breaking down barriers: between the humanities and the sciences, between clinical and public health medicine, between concepts of genetic and environmental determinants of health, and between the basic and clinical sciences.

A number of challenges lie ahead for public health genomics, both intellectual and practical. Intellectual challenges include the need for practitioners of public health genomics to have a broad understanding of all the fields of knowledge that feed into the enterprise: genomic science, the population sciences, and relevant insights from the arts, humanities and social sciences. At a practical level, there is a need for leadership in public health genomics: to develop general "genomic literacy" in the public health workforce, to build a cohort of specialists within public health and other professional groups who have a detailed understanding of the field, and to inform intelligent and evidence-based implementation of genome-based tests and interventions in health services.

Progress towards the goals of public health genomics can be accelerated by communication and collaboration. International networks such as GRaPH *Int* and the European group PHGEN (Public Health Genomics European Network)[12] will help to bring the vision of using genome-based knowledge and technology for the benefit of population health closer to reality.

FIGURE 1

A strategy for the effective translation of genome-based knowledge and technologies for the benefit of population health. The shaded parts of the diagram represent the components of the "enterprise" of public health genomics.

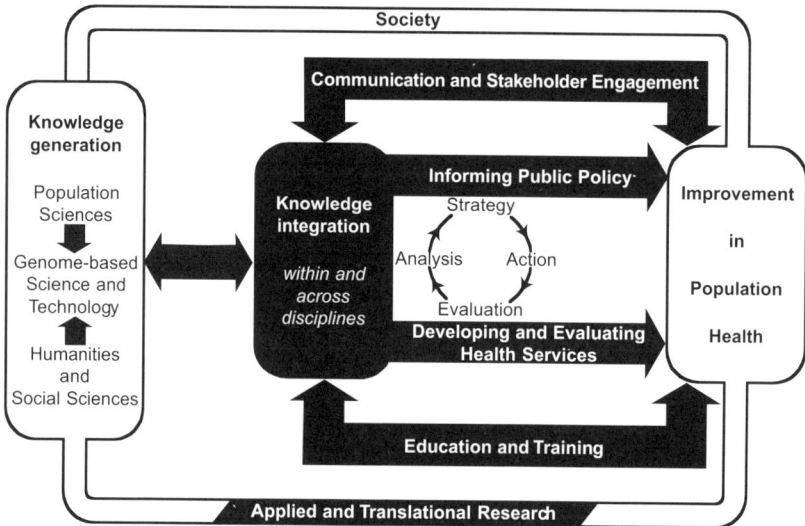

REFERENCES

1. M. Khoury, *From Genes to Public Health: Applications of Genetics in Disease Prevention*, 86 AM. J. PUBLIC HEALTH 1717 (1996).
2. Centers for Disease Control and Prevention, Translating Advances in Human Genetics into Public Health Action: A Strategic Plan, (2001), http://www.cdc.gov/genomics/about/strategic.htm.
3. National Office of Public Health Genomics (formerly the Office of Genomics & Disease Prevention), http://www.cdc.gov/genomics.
4. Institute for Public Health Genetics, University of Washington School of Public Health, http://depts.washington.edu/phgen. See also Public Health Genetics Interdepartmental Concentration, University of Michigan School of Public Health, http://www.sph.umich.edu/genetics.
5. NHS Central Research and Development Committee, The Genetics of Common Diseases (1995) and Genetics Advisory Group, Report of the Genetics Advisory Group (1995).
6. Public Health Genetics Unit, http://www.phgu.org.uk.

7. E. T. Juengst, *'Prevention' and the Goals of Genetic Medicine*, 6 HUM. GENE THER. 1595 (1995).

8. T. Murray, *Genetic Exceptionalism and 'Future Diaries': Is Genetic Information Different from Other Medical Information? in* GENETIC SECRETS: PROTECTING PRIVACY AND CONFIDENTIALITY IN THE GENETIC ERA, 60 (M. Rothstein ed., 1997).

9. W. Burke et al., *The Path from Genome-Based Research to Population Health: Development of an International Collaborative Public Health Genomics Initiative*, GENET. MED. (forthcoming 2006).

10. GRaPH Int website, http://www.graphint.org; Genome-Based Research and Population Health (Report of an expert workshop held at the Rockefeller Foundation Study and Conference Center, Bellagio, Italy, 14-20 April 2005), http://www.graphint.org/docs/BellagioReport230106.pdf.

11. *Id.*

12. PHGEN (Public Health Genomics European Network), http://www.phgen. nrw.de.

The Role of International Stakeholders Patients as Partners

Alastair KENT

Genetic Interest Group, UK
European Genetic Alliances Network

1. INTRODUCTION

Across the developed world, it is increasingly the case that the old linear, hierarchical relationship between doctors and patients is being replaced by a more balanced one in which patients are seen as partners, alongside other key stakeholders, in the process of addressing unmet medical needs and understanding the flaws in our biology that result in serious diseases that limit the quality, and often the duration, of life for those affected. Nowhere is this more evident than in the case of chronic diseases where, by definition, there is no intervention that will treat or cure the condition, and management is targeted at controlling symptoms and enabling those affected to enjoy the best possible quality of life, free from the worst effects of their condition. Although remarkable progress has been made in the control of some hitherto life-limiting diseases, it still remains the case that, for those affected by them, intervention is more often than not a case of the "least worst" rather than the "best possible" outcome.

As a consequence of the intractable nature of many chronic diseases, a number of important events have come together to make a powerful force for change—one that gives hope that many of today's unmet health needs will move from being incurable to being treatable, to being curable, and perhaps even to being preventable.

The factors that have brought about this shift, and opened up new possibilities for health gain for those with chronic disease include (*inter alia*):

- The advances in understanding molecular biology and our growing awareness of the relationship between our genes and our health in common complex disorders as well as in rare, single gene ones.
- Progress in informatics that enables vast amounts of data from different sources to be assembled and processed in meaningful ways.
- The willingness of patients and the public to bond together to commit to the research and to the support of the institutions and associations undertaking it—not blindly or through a naïve belief that "doctor knows best" but on the basis of an informed, subtle understanding of the risks and benefits to be had from collaborative working to address serious, unmet health needs.

It is this third issue that will be addressed in the remainder of this article.

2.   THE RISE OF PATIENT ALLIANCES

Across the developed world, the rise of support groups for patients with particular conditions has been followed by the emergence of thematic alliances at national, regional and global levels. Genetics is a field in which this is particularly the case, and in the last twenty years or so, we have seen the emergence in the United Kingdom of the Genetic Interest Group (GIG), with about 140 condition-specific support groups in membership; in the Netherlands, VSOP has about 70 members; and the Genetic Alliance in the USA has 600 member organisations. At a regional level, national patient alliances have grouped to form bodies such as the European Genetic Alliances Network (EGAN), which in turn contributed to the International Genetic Alliance (IGA). Nor is this a Northern Hemisphere phenomenon: it is repeated in Australia and South Africa, and groups are beginning to emerge in China and elsewhere in Asia.

What these groups have in common is the ability to intervene in a strategic way with policy makers, planners and the professional community. They present a co-ordinated, patient-focussed perspective on current hot topics in the continuum from research and development through implementation and adoption to reimbursement, with the ability to act as powerful advocates in the political and the public arenas, where issues relating to science and health care are under the microscope.

Thus, for example, GIG is currently working on issues to do with the secondary use of tissue samples for research, EGAN is lobbying the European Parliament about improvements to the proposed Regulations on Advanced Therapies and Tissues Engineered Products,[1] whilst the IGA is working with the World Health Organization (WHO) on raising the profile of avoidable birth defects in the poorest nations of the developing world.

3. BIG ISSUES, BIG SOLUTIONS

The cost of chronic complex diseases—cancer, heart disease, diabetes, mental illness, osteoporosis, etc—is huge. Phenotypically disparate, these conditions all share the fact that they arise from a combination of predisposing factors, genetic environmental and lifestyle, all of which interact to precipitate the emergence of the condition in the patient.

Attempts to understand the basic biology of these disorders has created the necessity for the establishment of large-scale population-based sample and data banks that will make possible the undertaking of sophisticated prospective studies that will hopefully result in the emergence of novel interventions for a whole range of conditions.

However, these biobanks often operate under conditions of considerable uncertainty, not least with regard to their financial viability over the timescale necessary to deliver significant results. Scientific time horizons of 10, 15 or even 20 years clash with politically-driven imperatives that focus on the end of the financial year or the next election, whichever is sooner. Ultimately, ongoing political support (and the continued financial investment that follows from this) is dependent on acceptance and endorsement by the public. The perception that large-scale data and sample banks are a vote loser will have politicians walking away from them very quickly, whilst their endorsement (and funding) will flow from the belief that these are important resources in the development of tools to combat serious diseases. Similarly, in the private sector, if the outputs of such ventures are seen to be unacceptable, offering a reduced prospect of a return on investment, then funds will dry up.

Patient engagement is an important tool for raising public and political awareness and securing support for the ongoing development of large scale biobanks. Unlike other stakeholders, all of whom might be seen to have a vested interest, patients are a disinterested (but not uninterested) advocate for them. Ultimately, patients do not care about the process, but they do care about the outcome because without high quality research and development, potentially treatable conditions will remain untreated and ill health and suffering will continue to blight the lives of those affected and those who care for them. If biobanks are the best route to achieve this, then they will get support from those who stand to benefit, thereby helping to persuade the wider public of the need for investment in these resources.

Patients are not just uncritical advocates for population-based research resources. They are also a source of information and expertise about the diseases which biobanks hope to help unravel, able to contribute to the development and use of these resources at all stages of their evolution, from the original "bright idea" to "pills in patients."

In the UK, for example, the Genetic Interest Group[2] contributed to the consultative process preceding the establishment of Biobank UK by organising events and meetings to elicit the views of a range of stakeholder groups, including patients with particular disorders, family doctors and members of the general public. GIG also undertook an independent analysis of public and professional perceptions and expectations of the UK Biobank, feeding this bank so that communication policy and strategy could evolve to reflect and respond to the issues that were identified. One of the important messages to come out of this consultative exercise was the extent to which the size of ventures such as Biobank UK, and the fact that they are funded by public sector bodies (the Department of Health and the Medical Research Council) as well as respected charitable foundations (The Wellcome Trust), makes them seem to be part of the statutory health care provision of the National Health Service (NHS). Indeed, the association of Biobank UK with the NHS "brand," which is possibly one of the best known brands in the United Kingdom, is an important component in securing and preserving the trust and confidence of the general public in the integrity and the viability of the venture. It is also important in helping individual citizens make up their minds whether or not they are prepared to volunteer to donate samples and allow their medical history to be included in this resource. Without the association in the public's mind of Biobank UK with the NHS, it is possible that the recruitment of volunteers might have proved difficult. Whilst no claims are made by Biobank UK in this respect, it undoubtedly benefits from the leap of association made by individuals when they learn of its aims and purpose.

Other groups, such as INVOLVE (formerly Consumers in NHS Research), have also been active advocates for patient engagement in all stages in the research and development (R&D) process. INVOLVE is a national advisory group, funded by the UK Department of Health, which aims to promote and support active public involvement in NHS, public health and social care research. It is governed by a board made up of patient representatives, academics and clinicians, all of whom are committed to this concept. Amongst its publications are: *Getting Involved in Research: A Guide for Consumers* and *Involving the Public in NHS, Public Health and Social Care Research: Briefing Notes for Researchers.*[3] A number of condition-specific patient groups have set up patient and family panels to sit alongside more traditional scientist and clinician-led review mechanisms to help steer research and establish priorities.

One such group is the Alzheimer's Society, which has established an advisory network of over 150 carers, former carers and people with dementia to inform the Society's "Quality Research in Dementia" Programme. Panel members play a full and active role alongside other stakeholders, helping to determine research priorities and providing comments and criticism on proposals submitted for funding. They also exercise an ongoing monitoring role and help raise awareness and disseminate results for research funded by the QRD programme and other relevant sources.[4]

A second British example is provided by the Motor Neurone Disease (MND) Society, which is investing over £1 million in setting up the MND DNA data bank. It will collect DNA, clinical and personal data from several thousand patients with MND, family members and controls over a five year period for use as a resource for researchers investigating causes and possible cures for this dreadful disease.[5]

The drive towards patient organisations taking a greater degree of involvement in research and in the setting up of databases and resources is not confined to the UK. Frustrated by the lack of interest in the condition that affected her child, a parent set up the European Network for Research on Alternating Hemiplegia (ENRAH), putting together a consortium of researchers in centres across the European Union, persuading the European Commission to fund the network, and in so doing, creating the critical mass to enable research to progress, and families and the clinicians who support them to be better informed about this condition and to provide the help that those affected need and can benefit from.[6]

Such panels, initially viewed with suspicion by some professionals who feared that basic research would suffer at the expense of projects nearer to the market, have come to be welcomed as they improve the quality and clarity of focus of proposals received, and encompass basic as well as near market applications. Patients, fundamentally, want cures. Only if a cure is unavailable (because we lack the knowledge to develop one) does palliation becomes the issue. Of course, if effective interventions are available, then they have to be delivered in timely, appropriate, user-friendly and equitable ways—which is why patients and families are also interested in Health Service Research.

4.   PATIENTS AS INVESTIGATORS

Although the theme for this article is "patients as partners," this should not be taken to imply that the patient is a follower rather than a leader in the process of developing sample and data banks for the investigation of the genetics of serious disease. There are a number of examples where such resources have been instigated and created by patients and families, who have also been responsible for securing the funding and determining the uses that are or are not permissible. Some of these resources may be condition-specific, but others are more wide-ranging in their coverage.

In the UK, a coalition of parents of children with Becker and Duchenne Muscular Dystrophy (Parent Project UK) have secured funding from the Department of Health and other sources (public, private and voluntary) to establish a sample and data bank for the investigation of this condition. The PPUK DMD Registry aims to gather data on every patient in the UK who is affected by either Duchenne or Becker Muscular Dystrophy in a secure, legally protected resource available to legitimate researchers and clinicians under agreed terms and conditions. It will facilitate basic research and the development of innovative interventions, including gene therapy and cell therapy, for patients with these conditions. Because it is "owned" by the families, issues relating to the consent and data-sharing have proved much more straightforward to address, and ethical approval for proposals more readily forthcoming than is sometimes the case with other sample banks.[7]

In the USA, the Genetic Alliance has set up a sample bank covering a range of rare conditions ranging from Inflammatory Breast Cancer to Noonan's syndrome, providing opportunities for the efficient use of the resources that would be individually too expensive and too demanding for patient groups to operate on their own, and providing a single point of access for researchers wishing to make use of the facility. A copy of the press release, taken from the Genetic Alliances website[8] is reproduced as Appendix 1.

Whilst many of the foregoing examples are physical collections of samples associated with datasets assembled in one place, this does not have to be the case. Using funding from the European Commission EURORDIS (the European Rare Disorders Association) has created a "virtual" biobank by building an overarching infrastructure known as EuroBioBank that links sample collections held in a number of different locations, creating a critical mass that will allow for better, more effective use of patients' samples and data—something which most patients see as crucial in the pursuit of novel interventions.[9] Again, the announcement of this resource is included as Appendix 2.

What is striking about these initiatives is the extent to which patient organisations have taken the initiative. They have recognised shortcomings in existing arrangements, created and helped to create the networks, built relationships, worked on research questions, joined in the design of infrastructures and operating protocols, secured funding (both directly and indirectly) and sometimes directly managed and controlled the use and development of the resource they have helped to bring into being. Nor does the story end with the understanding of reliant high quality research. Patient ownership (or at the very least their active partnership) in the management and operation of sample and data banks can also help address the issue of commercial development of research outcomes—often a thorny issue is the mind of the public, worried by the stereotype of rapacious pharmaceutical companies making apparently huge profits off the back of publicly-funded research and development. Rational resolutions to vexed questions such as how to create appropriate models for benefit sharing in this context are much easier (and likely to be more robust) if the ultimate end-users (the patients) are at the negotiating table alongside the other stakeholders. For example, Cancer Research UK used its ownership of the BRCA2 patent to enforce availability clauses in licences to develop diagnostic tests for UK patients, whilst Genzyme has actively collaborated with international patient alliances for people with the inborn errors of metabolism for which it produces therapies over issues such as compassionate use for its products in poorer countries unable to afford the market price and/or lacking the clinical infrastructure to deliver these safely and effectively.

5. VALUE FOR MONEY

Large-scale population genetics and genomics projects are expensive to establish and to run. It is in everybody's interest that they deliver efficiently, effectively, ethically and sustainably, meeting all appropriate standards with regard to the quality of their outputs and the protection of the interests of sample donors from undue risk or exploitation.

Patients have a particularly keen interest in seeing the maximum value (in terms of new knowledge and the possibility of innovative interventions emerging) being squeezed from the sample and datasets that Biobanks comprise. For this to happen, interoperability is essential. Whilst this is in many respects a technical issue, it is also intimately bound up with trust and confidence of the public and the sample donors in the integrity of the people and the robustness of the systems that control the uses of samples and data permitted, ensuring both that these are in line with the original consent given at the point of sample donation and that consent is given on the basis of the donors' understanding of the potential for unknowable future uses given the state of current knowledge. Again, patient and user participation in the creation of ethical frameworks, standard operating procedures and other control systems and decision-making processes will help ensure that they are appropriate, and sit alongside technical and scientific measures to protect the integrity of the resource, and ensure that appropriate, high quality scientific outputs can be produced reliably, effectively and efficiently—thereby securing the sustainability of the resource and the maintenance of public trust.

6.   CONCLUSION

The advent of large-scale sample and data banks has created powerful new tools for the investigation of many of the complex common diseases that currently cause untold distress, suffering and premature death throughout the world. Many of these banks are still in their infancy, but they offer hope of a better future for many. To realise their potential, they must become part of the science infrastructure—and be seen as a long-term resource, not a short-term quick fix. Sound science will be essential for this, but so will the maintenance of public and political confidence and trust in the robustness of the regulatory frameworks put in place to prevent abuse.

Active participation by patients as partners in the processes for establishing and governing these resources will help to ensure that these are appropriate and proportionate to the risks and benefits—nuanced by the reality of the experience of contractible health problems, avoiding the temptation to look for a "one size fits all" approach, yet capable of offering a robust protection for end users, scientists and clinicians alike which will help to ensure that the potential of these "big biology" resources is realised for the benefit of those currently waiting.

APPENDIX 1

GENETIC ALLIANCE BIOBANK LAUNCHED

Washington DC – October 27, 2004. Seven genetic advocacy organizations established the Genetic Alliance BioBank™, a repository for the standardized collection, storage and distribution of biological samples and clinical data for research purposes. This novel, advocacy-owned and -managed repository focuses and accelerates research, providing infrastructure for many advocacy groups to build a valuable resource. The Genetic Alliance BioBank™ sets new standards for participant involvement in research, provides standardized protocols, allows for ethical re-contact and robust protections in the context of the communities served by these advocacy organizations.

"Our organization has longed for the day when we can focus research with this resource, and create a dynamic consortium of researchers driving toward the same goal – accurate and timely diagnosis followed by effective treatment of inflammatory breast cancer," said founding board member Owen Johnson, President of the Inflammatory Breast Cancer Research Foundation. They founded the BioBank with six other organizations: CFC International, Joubert Syndrome Foundation, National Psoriasis Foundation, NBIA Disorders Association, Noonan Syndrome Support Group and PXE International.

The Genetic Alliance BioBank™ follows a model established in 1995 by PXE International. That rare disease organization has initiated and conducted research on pseudoxanthoma elasticum (PXE), actively participating in gene discovery and patenting, and development of a diagnostic. PXE International and the other groups came together through their work with the Genetic Alliance, a coalition of over 600 advocacy organizations. These founding members developed standardized model documents for the bank, designed and approved by the Genetic Alliance BioBank Institutional Review Board.

Researchers who wish to receive samples submit an application to the disease-specific advocacy organization. These organizations release coded samples to the researcher and hold the key that connects specific samples to individuals, offering a unique opportunity to enable follow-up studies while protecting participant confidentiality. The Genetic Alliance BioBank™ will help accelerate basic and translational research and serve as an essential platform solution for applying the tools of genetics, genomics, proteomics and metabolomics. The GA BioBank™ will also provide an opportunity for cross-disease research that may shed light on pathways and etiology for both common and rare diseases. The Genetic Alliance BioBank contracts with PreventionGenetics of Marshfield, WI for sample archiving.

"The BioBank is evidence of the next generation of patient advocacy," said GA BioBank™ founding President Sharon Terry, "But this is only the beginning. We are managing this resource, this community, with our eye on the prize – we will positively impact health outcomes. Solving these problems is often the work of generations, but we are taking one giant step in our lifetime."

APPENDIX 2

EUROBIOBANK: BIOBANKS FOR RARE DISEASE RESEARCH

Time has come for the leading European network of DNA, cell and tissue biobanks dedicated to rare diseases to take stock and reflect on their achievements. This 3-year project financed by the EC (2003-2005) has been highly ranked for its scientific value: thousands of samples distributed (approximately 6 800 in 2004) and, in August 2005, the online cataloguing of all the banking partners' collections on the EuroBioBank website (www.eurobiobank.org). So far, this catalogue includes 140 cell collections, 486 DNA collections and 287 tissue collections from rare disease patients. The network was established by patients and researchers with the aim to facilitate research on rare diseases by guaranteeing quick and easy access to samples via an online catalogue. This positive impact on European citizens was acknowledged and EuroBioBank was awarded the Newropeans Grand Prix 2004 for the best European project in the category Research & Technology.

When a researcher needs biological material, he/she only has to access the EBB website (Services>Catalogue of collections) and use the search engine to find the samples required. One click on the biobank's e-mail address next to the desired sample and a form appears. The researcher simply fills the form out and sends it to the biobank to obtain the samples necessary for his research project. This way, the biological material is exchanged much more quickly, thus speeding up rare disease research.

One of the tasks of the network is also to promote quality banking practices of collect, preparation, storage, and transport of biological material, and to address ethical issues relating to these practices. Again, remarkable progress has been achieved in this field: the EuroBioBank partners worked in common to develop harmonised Standard Operating Procedures (SOPs)1 and a Material Transfer Agreement (MTA)2 that comply with the OECD's recommendations for Biological Resource Centres (BRCs)3. These documents have been published on the EBB website and are now available to the scientific community.

Moreover, the survey conducted among the partners of the network on ethical issues resulted in the publication of an innovative book on the ethical and legal implications for biobanks. This book gives an overview of current legislation in the different member states represented at EuroBioBank.

A group of EBB partners is currently finalising the Network Charter, a document that governs the organisation of the network and the status of its future members. Currently composed of 12 BRCs representing 8 member states, the network aims, in the long-term, to expand with the addition of new BRCs and the development of partnerships with other networks, thus accelerating progress towards new therapies for approximately 30 million patients suffering from rare diseases in Europe.

In the past 3 years, EuroBioBank has been instrumental in increasing rare disease research. It is the only service infrastructure of this type for rare diseases. The long-term continuity of this network is a priority for rare disease patients; it is therefore necessary that Europe, and the national authorities, acknowledge the work done by EuroBioBank and support it in the future.

For more information about the EuroBioBank activities:

website:	www.eurobiobank.org
e-mail:	eurobiobank@eurobiobank.org
Appendices:	Appendix 1
	http://www.biobank.org/default.asp
	Appendix 2
	http://www.eurordis.org/article.php3?id_article=451
Websites:	www.enrah.net
	www.eurordis.org
	www.geneticalliance.org
	www.gig.org.uk
	www.invo.org.uk
	www.mndsaaociation.org/research/dna_bank/index.html
	www.ppuk.org
	www.qrd.alzheimers.org.uk/QRD_advisory_network.htm

REFERENCES

1.   Commission of the European Communities, *Proposal for a Regulation of the European Parliament and of the Council on Advanced Therapy Medicinal Products and Amending Directive 2001/83/EC and Regulation (EC) No 726/2004*, COM(2005) 567 final, Nov. 16, 2005, http://pharmacos .eudra.org/F2/advtherapies/docs/COM_2005_567_EN.pdf.

2.   Genetic Interest Group, http:// www.gig.org.uk.

3.   J. Royle et al., Getting Involved in Research: A Guide for Consumers (2001); B. Hanley et al., Involving the Public in NHS, Public Health and Social Care Research: Briefing Notes for Researchers (2004) ; *See* also www.invo.org.uk.

4.   QRD Programme, http://www.qrd.alzheimers.org.uk/QRD_advisory_network .htm.

5.   Motor Neurone Disease Association, DNA Bank, http://www.mndassao ciation.org/research/dna_bank/index.html.

6.   European Network for Research on Alternating Hemiplegia, http://www.enrah .net.

7.   Parent Project UK, http:// www.ppuk.org.

8.   Genetic Alliance, http://www.geneticalliance.org.

9.   European Rare Disorders Association, http://www.eurordis.org.

Genomics and Modes of Democratic Dialogue: An Analysis of Two Projects[*]

Hubert Doucet
Marianne Dion-Labrie
Céline Durand
Isabelle Ganache[**]

Groupe de recherche en bioéthique de l'Université de Montréal, Canada[***]

1. Introduction

Since the beginnings of bioethics in the mid-1960s, words such as "communication," "deliberation" and "dialogue" have become core concepts for expressing the task of ethics in the field of biomedicine.

In what we could consider the first phase of bioethics, these concepts were mainly used with the aim of obtaining informed consent, either from a human subject for research purposes or from a patient for treatment. In order for a lay person to understand the type of intervention the expert wanted to perform, the expert had the duty to enter into dialogue with him or her. There was a need to establish communication between the two. And after due deliberation, the lay person was invited to make her or his decision.

For at least three or four decades, this form of information transmission from an expert to a non-expert was considered the most perfect ethical tool for protecting human subjects in the field of biomedicine. Ethics consisted mainly of a dialogue between expert-beneficence and lay person-autonomy.

Such an approach to ethics has now become too limited, too restricted. It deals only with one side of the situation with regard to biomedicine, biotechnology, and especially genomics. Modern biology does not only help improve the state of health of sick people or help cure them, it also transforms our expectations with regard to medicine and biotechnology. We could even go further and say that modern biology causes profound changes in our whole way of life and thought. The issues are no longer of an individual nature or arising between two individuals. They are social, economic and cultural. Is the world built by modern biology, the world we want to live in? Science can no longer be separated from politics, in the most noble meaning of the word.

This new context helps understand why concepts like "dialogue," "communication," and "participation" take on a new dimension, a collective or public one. In the field of bioethics, we speak now of citizen dialogue. Recently, new approaches have seen important developments. They are promoted by various organizations, as much by technology evaluation agencies as by national ethics committees. These new approaches aim at "involving the citizenry in the decisions that affect them."[1] Alongside representative democracy thus takes place what we could call participatory democracy.

Since 2001, GREB (the *Groupe de recherche en bioéthique*) has devoted its energy to studying and developing means of promoting public participation in decisions to be made in the field of genomics. Within a large interuniversity research project called *Genomics in Society: Responsibilities and Rights* funded by Genome Canada and Genome Quebec, GREB was responsible for the communication platform. To carry out its responsibility, it has put in place a *Citizens Forum* which included two different activities. First, theatre: with the support of theatre professionals, we created a play on genomic issues, a play with the goal of facilitating public dialogue. The play has been presented to diverse audiences (the general public, scientists, college and university students, etc). Second, a "citizens conference," which took place in February 2005.

Our ethics research group has introduced an innovation in the sense that ethics groups usually examine or evaluate work done by others or recommend types of work that could be done. In our case, we initiated the projects themselves. By launching these activities, our hope was first to increase public interest for such types of work. Also, by critically examining our own activities dealing with public deliberation, we wanted to highlight basic conditions for a real public dialogue or a true citizen participation in our own context. In this day and age, when ethics is quite fashionable, it is very important to avoid misrepresentations.

2.   THE CITIZENS CONFERENCE

The evaluation of new technologies presents major challenges to modern society. For centuries, governments have used commissions or hearings to invite the public to discuss important issues or to learn their views about new technological developments. With the complexity of choices and the multiplicity of actors, other mechanisms have been put into place to integrate citizens into these reflections. New approaches have recently emerged, such as consensus conferences, citizen conferences, and citizen juries.[2] Most of the time, these tools bring together about fifteen citizens, non-experts chosen randomly. The citizens are first trained on the topic of the conference. The training must be as objective and well-balanced as possible. Indeed, it should introduce the participants to different perspectives and trends, in order to avoid biases. The goal of the preparatory phase of the conference is to prepare the citizens to discuss with the experts whom they will meet during the public phase of the conference. At the end of this first phase, the citizens, by consensus, will deliberate and make their recommendations public before they are taken into consideration by the authorities. These mechanisms allow lay people to establish a genuine dialogue with experts on complex scientific questions that concern them as well as on their social impacts. The general public is usually invited to participate.[3]

Entitled « Et l'Homme créa la génomique! » ("And Man created genomics!"), with the subtitle « Les avancées de la biologie humaine à l'ère de la génomique » ("Advances in human biology in the era of genomics"), GREB's citizens conference adapts the traditional formula of Denmark's consensus conferences and of other citizen conferences to reflect on the reality of genomics in Quebec and promote active participation by the audience. The citizens met over two weekends in the fall of 2004 to prepare themselves for their public session with the experts during the first weekend of February 2005 and produced a public report on their experience.[4] Our goal in realizing this citizens conference was to create a real dialogue between lay people and experts about genomics. We based our approach on the conviction that every individual has life experience that deserves to be heard, that each perspective could help clarify a subject and bring new elements of reflection, and that an exchange of values between communities is possible. Reaching a general consensus in the comments and recommendations of the citizens was not the goal.

A year later, the qualitative evaluation of this conference as a mechanism of citizen communication allowed us to confirm our three basic premises:

- citizens are capable of understanding science and passing judgement on it;[5]
- experts are able to enter into dialogue with society-at-large; and
- science must operate within the limits of the democratic will.

This evaluation consisted of a qualitative analysis, with the help of the N'Vivo software, of 14 interviews with participants in the citizens' conference (there were 36 participants total), of 24 questionnaires filled out by members of the audience and of the citizens' report.

The present evaluation emphasizes four points: the relevance of the mechanism, its contribution, the requirements raised by different categories of participants, and the process itself. More specifically, the conference participants considered it a good mechanism of citizen communication and appreciated their experience. The strength of this activity lies, indisputably, in the meeting, the exchange between citizens (members of the citizen panel and the audience), and experts. "Rare are the moments when we have the opportunity to exchange with so many amazing people," commented one expert. This exchange illustrates two-way communication, the theoretical foundation of the *Citizens Forum on Genomics*.[6] In such a conference, experts must be willing to initiate a dialogue with the public; they should not act as professors who limit their task to transmitting information to their students. They must behave as citizens discussing with other citizens. In most cases, this is what they did. According to some participants, however, one or two experts did not display interest in the dialogue.

In our evaluation, we noted that experts were very impressed with the type of exchange they had with the citizens. "I would participate in other citizen conferences with delight, because these exchanges with the public were very rewarding for me," one of them said. Before the conference, some scientists expressed their fear that the discussion might turn into a fight. At the conference, they recognized that the participants had shown a remarkable type of "popular wisdom." The citizens knew the issues related to genomics, what was good or not good for society. The fears did not come from ignorance on the part of citizens, but rather from a certain wisdom, according to the experts and the conference moderator.

> "On this subject, it is essential to enter into dialogue with experts from other disciplines and especially with citizens, not only to inform the latter but above all to listen to them and benefit from their wisdom and good sense."

> "The people have 'popular wisdom.' There is a knowledge, a tacit knowledge among citizens that is very strong, and what was quite formidable was to see how the discussion, the exchange allows these elements to emerge that create a climate of confidence, of respect, of listening. We saw coming from different participants with different points of view the beginning of very interesting questions, fundamental questions, in a certain sense."

Does this mean that if scientists would give more information to the public, they would easily convince the public that their projects should always be given priority over all other elements of our social life? No, citizens have an inner capacity to identify values and limits related to genomics and to indicate the best choices they consider important to promote, according to the common good. Also, the participants at this conference appreciated the open-mindedness of the experts. "The experts proved themselves to be open and even desirous of exchanging with citizens about the issues related to their work." They were ready to go. However, a weakness of these exchanges lay in the lack of time. The experts felt constrained. "I had not had time to nuance [my position] in my presentation because I really had limited time to give my arguments, on which my presentation was based."

The citizens conference had an impact on the expert participants in various ways. The first follows the exchanges that took place with citizens during the public citizens conference. In particular, the panel of experts discovered the citizens' competence in exchanging, in expressing opinions on genomics. "What filled me with enthusiasm in this project was to see to what point the citizens present during the weekend were well-informed and had a sharp critical sense regarding developments in genomics." The experts discovered the "citizens" vision, a considerable asset to their work. Finally, the citizens conference allowed the researchers to leave their laboratories. "That is another strength, that is, taking researchers out of research environments," according to one expert.

For the citizen participants, this conference was a great opportunity to increase not only their knowledge about genomics but also to develop a better understanding of science, what we would call scientific culture. The citizens are better informed, more demanding in terms of information. They realize the complexity of genomics and discovered the reality of research. According to one of the scientific experts, "afterwards, people could possibly be less naïve in terms of the information they receive and could become more critical. That is pretty important." At the end of the adventure, they had a new feeling of responsibility with regard to science, especially genomics. Citizens must take part in the public debate, in the dialogue on scientific advances and even in decision-making. For the citizen panel, "the voice of citizens could be stimulating for experts but also a significant support for their work before political and economic decision-makers if the dialogue that emerges is productive." Citizens have a role to play, a responsibility towards scientific development.

"Generally speaking, participants on the panel believe that citizens have a role to play in public debates about genomics. Citizens, in expressing their concerns and lack of knowledge about this new science, will allow a society-wide debate on questions such as: how far are we ready to go?"

More specifically, citizen participation allows us, according to one expert, to develop a common vision on science, to work for the interests of all.

"In effect, a frank discussion, involving us all, is essential not only in order to share our perceptions and preoccupations, but especially to develop a common vision of the objectives to achieve in the short- and medium-term, in order to ensure a harmonious and respectful transition of the interests of citizens."

On the other hand, two citizens expressed doubts about the actual influence of citizens. In effect, the goal of the citizens conference was not to have citizens participate in a decision-making process but rather to create a dialogue between citizens and to experiment with a new mechanism of citizen communication. Finally, citizen participation is demanding. Citizens must be available for several meetings, be capable of understanding and deepening information that is sometimes abstract, be readily able to exchange ideas, to ask questions of the experts, to listen to other citizens and, most importantly, to believe in the process, the ability and the power of citizens.

Experts also have a role to play regarding the democratization of science. They must leave their laboratories and join the public.

"They must leave their universities and their laboratories and go to public places, to participate more in public conferences, to reach decisions at events that could be also oriented towards the development of science," according to one expert.

The evaluation of the citizens conference is also concerned with the process. Participants experimented with a mechanism of participatory democracy that allows or the democratization of science, of decision-making in genomics, and the involvement of citizens. "It is a means of giving them information, as I said, and a means that is democratic" stated one expert. Citizens have real power in the realization of the process. This fact is important for the advisory committee, which ensures, with the organizers, the neutrality and objectivity of the activity.

> "It is really a mechanism of participation and the impact that it has, I find, on people who are deprived of power in the beginning, of feeling that they can say something, that they can learn something, and then that they can participate and everything, it's that that changes individuals."

On the other hand, the whole process demands a lot of energy for all parties involved. Some people have wondered whether such a process is not too demanding for the result it produces. The conference certainly allowed a useful exchange between experts and citizens, even if it is not a decision-making process. The participants as a group recommend holding a citizens conference with a specific purpose in order for it to be in line with, or for it to take place in parallel with, a specific event like the elaboration of policies regarding genomics. It seemed, for the moderator, to be an important element. "Even when the objective is so ambitious and the mechanism so demanding as a citizens conference, it seems to me that it must be integrated into a process that leads to something."

The citizens conference must also be repeated. It is even part of scientific responsibility, according to one expert.

> "The exchanges between representatives of the public and of the scientific world, having to do with issues that come from genetic biotechnology, must be made a part of the scientific responsibility, especially when research funding comes from public funds. Consequently, it is imperative to organize places to exchange ideas and enriching initiatives like the citizens forum because they fit into exactly into this mold."

The dissemination of this activity must also be increased. Several participants were disappointed that participation from the public came in limited numbers to the conference (one hundred or so members of the audience for the public weekend).

To conclude this section about the citizens conference, two elements deserve to be emphasized. This type of mechanism necessarily raises the question of the representativeness of the participants. Some members of the citizen panel considered the lack of representativeness a weak point and an element to be improved. One citizen commented:

"I had been extremely...annoyed...I must tell you, annoyed not more than that but annoyed from the beginning by the, the representation of citizens around the table and of course we don't have control over that but I had the feeling that we were, that we, that we represented ... we represented, like, a part of society, a part maybe relatively educated, relatively calm maybe, in its values."

Maybe there was miscommunication between the citizen panel and GREB about this question. For the latter, it was clear that a panel of citizens cannot be representative of the public. In preparatory meetings with many speakers, we had been widely questioned on this aspect. We always responded the same way. A citizens conference does not pretend to be a process of representativeness. Its critical element is to "allow individuals with different backgrounds, interests and values to listen, understand, potentially persuade and ultimately come to a more reasoned, informed and public-spirited decision."[7] Before another conference, this point would have to be more clearly explained.

The evaluation of this citizens conference also brings to light the necessity of developing different mechanisms of citizen communication, not just a single one. Always using the same mechanism might create a democratic deficit in society, from which the importance of developing different mechanisms stems. This is what GREB did in undertaking a second activity of citizen communication; namely, a piece of interactive theatre about advances in human biology in the era of genomics.

3.   THE PLAY: THE THEATRE OF SOCIETY[8]

Why did we present a play in parallel with the citizens conference? Simply because theatre is an art that is able to transmit not only ideas, by facilitating new knowledge, but also emotions. It appeals to each human being in his or her totality, as much at the affective level as at the intellectual level, by creating an opportunity for rich communication. In building their democracy, the Greeks understood that theatre was the making of an assembly (a gathering) where, on the one hand, topics of great depth could be tackled and, on the other hand, a distance remained with the drama or the tragedy. In a way, theatre mixes business with pleasure. It is open to everyone and facilitates discussion and public reflection.

Theatre is another means of promoting public dialogue and of promoting it differently. This endeavour had four goals:

- to arouse the interest of citizens for a new world, the world of genomics;
- to feed a social reflection on the challenges raised by advances of human biology in the time of genomics;
- to open a respectful and creative space for dialogue about the stakes in genomics; and
- to facilitate a large public debate about the challenges raised by genomics by gathering people with different perspectives using an innovative and original means.

The play was called "The Theatre of Society." It was written by a group which specializes in popular theatre. In order to become acquainted with the topic of genomics and remain open to the various viewpoints on the topic, the author himself had to enter into dialogue with a large number of people. The interactive aspect of the play facilitates dialogue with the audience. The play is divided into five scenes, each discussing a different theme in relation to the advances of human biology in the time of genomics. The production has been performed fives times before different groups (researchers, college and university students, interested members of the public, etc.) In order to collect the participants experience, to know what they drew from it and what they think of the form and the content of the piece, questionnaires were filled out (by 28% of participants), a discussion workshop with the vast majority of spectators took place after each performance, and semi-directed interviews took place with the creators of the play. In general, the experience of interactive theatre seemed to be appreciated by the participants as well as by the creators of the piece, as demonstrated in the table below.

TABLE 1: SUMMARY TABLE OF QUANTITATIVE DATA

Appreciation of the play	87%
Information about genomics	64%
Discussion of issues	74%
Presentation of a diversity of points of view	80%
Advancement of reflection	70%
Feeling of being involved in the issues	80%

"I very much appreciated the use of theatre with the goal of illustrating the current situation of the issues caused by genomics." Many participants said that they enjoyed it, that they were entertained, or that they found the experience pleasant, interesting, enriching, or educational. Moreover, the integration of humour into science was extremely appreciated.

The interest in theatre as a mechanism of citizen participation has various characteristics, as emphasized by the participants. There is, on one hand, the piece itself and, on the other, the discussion that follows. In terms of the piece, five strengths deserve to be mentioned. First, theatre constitutes a useful means of communicating information: it requires the science to be rendered accessible to the general public, which allows us to easily assimilate the subject despite its complexity. "Light words to help with a heavy subject." Then, the presentation of issues in the form of situations that could be lived gives participants the feeling of being concerned or affected, thereby favouring participation. "Situations that affect us, anchored in reality." The interactive aspect of the piece is also a strength: it allows us to maintain the spectators' interest, to stimulate the intellect and promote emotional participation, thus making a balance with the rational. "The rational side and the emotional side create the beauty of the human being, especially if there is equilibrium between the two." Finally, theatre offers the possibility of reaching a public of all ages and constitutes a good means of democratizing science as well as of reaching citizens without the intervention of the media.

Based on the discussion, two strong points received attention. The discussion that followed the piece allowed various points of view to be exposed in an open and respectful atmosphere. "It is an excellent means; people are open after the presentation and respectful." Participants consider it an opportunity to be able to exchange (to express themselves and to listen), allowing them to open a dialogue and deepen their reflection. "It is a chance to discuss our opinions and to hear others."

The evaluation of this mechanism of public participation demonstrates several weaknesses the participants emphasized, particularly the college students. According to this group, the information transmitted was insufficient or irrelevant, lacking specificity about the current state of genomics. "Seeing as we will soon arrive on the job market, we would have to have specifics about what they really do in this field and in these situations." It is important to note that these students attended the piece as part of a course, therefore with the intention of receiving "concrete" lessons more than to have a reflection or process of questioning proposed to them. Regretting the little time accorded to experts, the college students would have liked to hear their opinions regarding current research as well as on the state of law. "However, I think that experts must take a greater role and explain to us the current scientific situation more than to leave the field open to students' opinions." The students knew that experts were present; they therefore expected that the experts would actively participate and directly inform them. A final point raised by all categories of participants relates to the difficulty of expressing one's personal opinion in public, in the fear of being judged or of displeasing other spectators. "I really liked this method of communication, but it was still difficult to speak in front of people and to state opinions, knowing that they could be upset by them."

The creators agreed to collaborate with GREB in the theatre project for different reasons and/or interests. First, the subject that was not well-known to them constituted a challenge, making the stages of research, learning, and meeting with experts an enriching process of apprenticeship. Moreover, the element of rendering science accessible interested them very much, just like sensitizing the public to issues related to genomics, because of the social reflection that theatre could bring.

> "We have in our mission and what we have done in Quebec and also outside is sensitizing people to various issues that occur in our everyday lives, genomics. Good, a bit like we say in the piece, it is something that is…that people don't know enough, don't talk about a lot related to the impact that this could have on the daily life of many people."

Another motivation was the specific context in which the piece had to be performed: a more global activity, complementary to the citizens meeting about ethical reflections on the subject. Unfortunately, the combination of the two activities could not be realized.

How did the play's creators evaluate the experience of theatre as a mechanism of public participation? One of the strengths of the experience relates to the preparation of the piece: the collaboration with GREB, the support at the level of research as well as the meetings with experts bringing different points of view.

> "It's really a very precious jewel. Access to people like all the specialists that we saw and   even people who brought very different points of view even sometimes that confronted each other but that's good for theatre, opposing points of view, they're good."

The establishment of the piece of theatre was realized with the help of scientists who met and discussed with the creators in order to contribute to increased knowledge, to enrich the reflection that contributes to the process of theatre creation. The creators regretted therefore the absence of citizen consultation during the process of writing in order to put in place situations that affect everyone. "We maybe under-estimated a little the importance of going to validate the points that we retained with ordinary people, members of the public that we would meet afterwards." Another weakness was to not have returned to consult the experts following the first stage of writing. That would have allowed them to advance the reflection and to further push the ethical questioning.

Participants in the play discovered many facets of genomics. Some became aware that they were under-informed. "I realized that I was very badly informed about research in general, and that I seemed like a spectator but I am also involved." The play served as a spark to begin their reflection on different dilemmas. The complexity of the issues and the diversity of opinions that they raise was a new discovery for many, like the difficulty of making choices as well as the consequences and responsibility that result from them. "The complexity of consequences and responsibilities *vis-à-vis* certain choices."

Several aspects were raised during the discussion or in the participants' responses in the evaluation. We will first discuss the current organization of research, a subject that was addressed from different angles: funding of research either by public or private sources, patents, and international competition. The political aspect of science was largely questioned by the public. Exchanges on the decision to abort or not abort a foetus suffering from trisomy (Down's syndrome), a topic the play addressed, also raised lively debates on eugenics. It helped show the difficulty participants have in taking a position regarding the choices to make, both in the play and in relation to scientific advances in the era of genomics. Finally, exchanges related to the effectiveness of theatre as a mechanism of public participation; two points received attention. On one hand, we highlighted the importance of the emotional side as well as the rational side and the way to begin with these to reach different publics. On the other hand, we raised the importance of presenting nuanced situations, where everything is not simply black and white.

The fears that citizens have regarding genomics could be summarized as follows: these advances will be used irresponsibly. This irresponsibility could take various forms. The common good could be forgotten, to the detriment of other interests, represents one of the inappropriate uses of the technology. Irreparable harms are also possible and raise concern. "To begin irreversible processes in which the real consequences won't be known for many years, and then, too late!" The same goes for the commercial dimension, which leads to commodifying the human body or to imagining these advances only in terms of money. "Abuse of power and commodification." Moreover, participants associate research for profit with the loss of human values, to the denaturalization of the human being, to the distancing from what is natural and losing sight of human needs. "For sure! In the long term, a denaturation of the human being." The question of power also generated fears that some people would like to transform and control everyone around them, thus disturbing the natural process. "What scares me is that we are modifying nature in relation to humans." Fears of discrimination, eugenics, and the loss or lessening of human diversity were also mentioned. Finally, some participants feared that decisions will be taken against public opinion, without consulting the population or without a societal debate. "Yes, if the questioning and issues are not subject [to public discussion] and that decisions will be taken by the scientific minority."

How can we address these concerns? In promoting a democratic process that would regulate research funding and that would privilege quality and transparency of information. "That there be continuous information for the public." The need for communication between the population and the scientific world is strongly desired: "places for dialogue in the context of diversity." A greater control of research is also desired: that limits be imposed, rules and multidisciplinary control. At the same time that we want to pursue research and developments in genomics, we talk about proceeding with caution. This caution is essential if we want results that benefit all. "Yes, I would like advances to be accomplished rapidly, but cautiously."

Theatre seems therefore to have allowed us to open a respectful and creative space for listening and speaking. A diversity of perspectives was heard, and people were able to express their opinions and to listen to others with respect, allowing them to deepen or to advance their reflections. "Having reflected, having been able to express myself and hear others very different and varied opinions during the exchanges." It is, however, difficult to know if dialogue was realized between participants since they spoke more of debate, exchange, and discussion but very little of dialogue.

4.   CONCLUSION

The *Citizens Forum on the Genome* and the two activities that it created demonstrate the interest in giving a voice to citizens from all situations, scientists as well as non-scientists.[9] For the participants as a group, to promote debate around scientific decisions is to democratize science and to answer a genuine expectation. As one expert noted, it is necessary "to find a way to collectively make science more accessible and to [ensure] a more democratic participation in decisions that are made." To do this, mechanisms of citizen communication, diverse, complementary and respectful of societal values, must be put into place in order to discuss choices to be made with regards to scientific development.

This idea of complementarity is present in all the evaluations of activities of the *Citizens Forum on the Genome*. Always using the same mechanism does not constitute a cure-all to problems of participatory democracy in science. "It is necessary to exploit several at the same time," says one expert. Similarly, the work of definition must begin by distinguishing the different existing mechanisms. The specific goals and objectives must also underlie the choice of a particular mechanism. The diversification of mechanisms of citizen communication and their careful use will prove significant to really involve citizens in scientific decision-making and to contribute effectively to the public debate about genomics.

REFERENCES

*.   These two research projects were made possible through funding from Genome
        Canada and Genome Quebec, in collaboration with the *Centre des sciences du
        Vieux-Port de Montréal*. For the citizens conference, we were also supported by
        the *Bureau de la biotechnologie et de la science de Santé Canada*, ECOGENE-
        21, the International Institute of Research in Ethics and Biomedicine (IIREB),
        the Canada Research Chair in Law and Medicine, the Institute of Genetics of
        the Canadian Institutes of Health Research (CIHR), the *Faculté des Études
        Supérieures* of Université de Montréal and the Canadian Commission for
        UNESCO. For the play, we were supported by the *Théâtre Parminou*, who
        wrote the manuscript, members of the Quebec's GE³LS network (Genomics:
        Ethics, Environment, Economy, Law, and Society), the *Centre de Recherche en
        Éthique de l'Université de Montréal* (CRÉUM), the Canada Research Chair in
        Law and Medicine, the Faculté des Études Supérieures of Université de
        Montréal, the Faculty of Arts and Sciences of Université de Montréal and *the
        Fédération des Associations Étudiantes du campus de l'Université de Montréal*
        (FAÉCUM).

**.   Hubert Doucet is Director of the *Groupe de recherche en bioéthique* (GREB) of
        Université de Montréal, Marianne Dion-Labrie is a research associate at GREB,
        responsible for the citizen conference on genomics; Céline Durand is a research
        assistant at GREB, responsible for the interactive theatre piece on genomics;
        Isabelle Ganache is the coordinator of GREB (2004 to 2006).

***.   The GREB team (2001-2006) responsible for the realization of projects was
        made up of Hubert Doucet, PhD, Research Director of GREB; Béatrice Godard,
        PhD, researcher and Director of the *Programmes de bioéthique* of Université de
        Montréal; Danielle Laudy, PhD, researcher and teaching counsellor at the
        Faculty of Medicine of Université de Montréal; Guy Jobin, PhD, researcher at
        GREB (2001-2002); Isabelle Ganache, PhD candidate, research coordinator
        (2004-2006); Marianne Dion-Labrie, PhD candidate, research associate,
        responsible for the citizen conference; Isabelle Gareau, MA, research associate,
        responsible for the theatre project (2001-2005); Céline Durand, research
        assistant, responsible for the theatre project (2005-2006); Éric Racine, PhD,
        collaborator, director of the neuroethics research unit of the *Institut de
        recherches cliniques de Montréal* (IRCM); and Guillaume Paré, research
        assistant.

1.   J. Abelson et al., *Deliberations about Deliberative Methods: Issues in the
        Design and Evaluation of Public Participation Processes,* 57 SOC. SCI. MED.
        239 (2003).

2.   J. Durant, *An Experiment in Democracy, in* Public Participation in Science: The
        Role of Consensus Conference in Europe (J. Durant & S. Joss eds., 1995);
        Commission de l'éthique de la science et de la technologie, Les enjeux éthiques
        des banques d'information génétiques: pour un encadrement démocratique et
        responsable (see La conférence de citoyens) (2003); G. Smith & C. Wales,
        Citizen's Juries and Deliberative Democracy, 48 POLITICAL STUDIES 51 (2000).

3.   Commission de l'éthique de la science et de la technologie, *supra* note 2; J.
        Grundahl, *The Danish Consensus Conference Model, in* Public Participation in

Science: The Role of Consensus Conference in Europe (J. Durant & S. Joss eds., 1995); Durant, *supra* note 2.

4. Panel citoyen, *Et l'Homme créa la génomique! Rapport citoyen de la Conférence citoyenne sur les avancées de la biologie humaine à l'ère de la génomique* (2005), http://www.fes.umontreal.ca/bioethique/GREB_PROD/ind ex_fichiers/Rapport%20citoyen.pdf.

5. T. Leroux et al., *An Overview of Public Consultation Mechanisms Developed to Address the Ethical and Social Issue Raised by Biotechnology*, 21 J. CONS. POL. 445 (1998); A. Davison et al., *Problematic Publics: A critical Review of Surveys of Public Attitudes to Biotechnology*, 22 SCI. TECHNOL. HUMAN VALUES 317 (1997).

6. J. Fawkes, *Public Relation and Communication*, *in* The Public Relations Handbook (A. Theaker ed., 2001).

7. Abelson, *supra* note 1.

8. In French the play was entitled *Un jeu de société*.

9. Groupe de recherche en bioéthique de l'Université de Montréal, Le Forum citoyen sur le génome (2003), http://www.fes.umontreal.ca/bioethique /GREB_PROD/index_ fichiers/Page521.htm.

Meeting of Minds: A European Citizen's Deliberation on Emerging Technologies

Marie-Hélène MOUNEYRAT

Comité Consultatif National d'Éthique (CCNE), France

The international citizen's forum *Meeting of the Minds* was interesting for various reasons. First, experts have much to learn from lay people, who bring new perspectives to what are familiar issues to those working in the field. Their views are frequently disturbing, and I believe we need to be disturbed. Second, it is clear that nowadays we must work on the international plane, with international exchanges on all issues. In this world, a uniquely national vision makes less and less sense. Finally, the main reason this forum was so interesting is that the progress of science, especially progress in particularly complex and controversial technologies like genomics, requires social acceptance to proceed.

Democracy exists in the field of scientific progress—but even things that are clearly desirable are not necessarily easy.

For that reason, method is very important. Examining questions of method will be my first point in this chapter. Then I will present the positive aspects of *Meeting of Minds*, as well as the problems and the negative aspects of the experience. As a conclusion, I will discuss the follow-up to this experience and gesture towards the future of citizen engagement.

1. METHODOLOGY

The idea behind the *Meeting of Minds* project was, for the first time, to gather together citizens from across Europe and allow them to engage with experts, stakeholders and, most importantly, with each other as they compared their views on the impact of developments in brain science on their lives. In a word, we wanted to discover what was desirable and what was not desirable, from their point of view.

The *Meeting of Minds* panel was composed of 126 citizens from nine countries: Belgium, Denmark, France, Germany, Greece, Hungary, Italy, the Netherlands and the United Kingdom. In each country, fourteen citizens were chosen at random, but nevertheless selected to include a broad range of age, gender and professions. For example, in the French delegation, there were seven men and seven women; the age range was approximately twenty to seventy; and participants included students, teachers, architects, stay-at-home mothers, etc.

The *Meeting of Minds* process began in May 2005 and culminated with the European convention of January 2006. There were two levels of work. The national level began first, spanning March to May 2005. This dialogue included meetings and debates between citizens, including discussions with experts and stakeholders. In France, these generally took place over a weekend in Paris. The second level was the European convention, where citizens from different countries discussed their respective national work; this meeting took place in Brussels.

In order to enable a true dialogue, across language and cultural barriers, the Convention used very innovative methods and employed technology to support the citizens in their deliberations. Professionally facilitated discussions around small tables ensured that each and every panellist had a voice. Every table communicated its results in real time. At key points during the weekend, the panellists used electronic keypads to vote on results which were then synthesized. Interpreters at the tables ensured that at each stage of the process, every panellist could participate fully, regardless of which language predominated at that moment.

At the end of the meeting, six of the most important issues, not specific to brain science, were identified as the basis for discussions. They are:

1.   Regulation and control
2.   Normality versus diversity
3.   Public information
4.   Pressures from economic interests
5.   Equality in access to care
6.   Freedom of choice

These six issues could also be relevant in the field of genomics.

After this first European convention, the third step was to return to national discussions. At the end of these discussions, each country produced a national assessment report. This stage occurred between August and November 2006.

The final step was the second European convention, which also took place in Brussels, at the end of January 2006. At the second Convention, delegates produced a set of European recommendations.

The design of the second Convention varied slightly from the first convention, but the basic features remained. Professional facilitation ensured the widest-ranging dialogue possible among the participants. We also included innovative dialogue formats such as "Carousels" of approximately 40 citizens representing the nine countries and also a so-called "European café" setting, at which citizens were able to rotate to the two other "Carousels" to learn about what has been discussed in other groups and provide their input and feedback. Throughout the Convention, a group of citizens and writer-editors worked together to draft the interim and final results of the discussions. During a final plenary, amendments were introduced and votes took place to finalize the recommendations that were the product of the work of the weekend. At the conclusion of the

Convention, on January 23rd, 2006, this report was delivered to the European Parliament and handed over high level European officials and representatives of policy makers, as well as to the European scientific and research community.

This experience had both positive and negative aspects that lead to various conclusions.

2.   THE BENEFITS

The first clearly positive result of this experience is the great and deep involvement of citizens in questions raised by the scientific progress in emerging technologies in biomedical research. It revealed both the moral imperative and the practical value of involving citizens in discussions on ethical questions raised by scientific progress and its applications. Many saw the *Meeting of Minds* mechanisms as a model for citizen participation and suggested similar processes both on the European and national levels in the future. Many also believe that more significant participation by lay people in scientific discourse could help to communicate complex results to a broader audience.

Interestingly, by the end of the experience, the citizens felt that they were actors in a real democracy and that they could play a real role in the scientific choices.

From the point of view of experts and stakeholders, it was very interesting to see that, in fact, the point of view of citizens regarding sciences was really a combination of fear, on the one hand, and hope, on the other.

Another interesting aspect was the European dimension of this experience. It demonstrated both that a European dialogue was possible to organize and also that there are significant differences between the different European countries. Many citizens were surprised at this heterogeneity, but most were loathe to admit it. It seems they prefer a European normality as opposed to a European diversity, and I think that this reality, though anecdotal, is quite troubling.

This brings me to the limitations of this exercise:

3. THE LIMITS

One limit is linked to the fact that, very often, citizens' reasons for participating in such a debate are personal; in other words, the citizens who participate often have a friend or family member who is affected by the issue under discussion. This family or friendly context can very much influence their position and the questions they ask, and this reality can certainly contribute to a more limited field of discussions. Patient engagement is one thing, but citizen engagement is something else: all citizens are potential patients.

A second limit lies in the difficulty citizens have in raising specific questions in connection to the topic under discussion. For example, during *Meeting of Minds*, the six main issues identified by citizens are not specific to brain science. Issues such as informed consent, equal access to care, etc, are applicable to other fields as well. This last limit brings me to one of the main problems: namely, the problem of the information the citizens use to develop their positions and, as a corollary, the role of experts and stakeholders in sharing information. It is possible that the lack of specific questions raised by participants reveals a lack of information in the lay community about some complex and controversial emerging technologies. The role of experts in the informing process is crucial. The question remains, however, when and how is it appropriate for them to intervene? There are two possibilities:

1. we can leave citizens completely free and independent, and the dialogue with the experts can take place only in a secondary step. The risk with this possibility is that the lack of appropriate information does not necessarily permit citizens to address all appropriate questions.

2.   The second process—which was chosen for the *Meeting of Minds*—is to organise the dialogue with experts and stakeholders *before* the citizen's discussion. The risk here is evident: there is the possibility of influencing their point of view, in favour or against the scientific progress. The honesty of the experts must also be considered. For example, at the second European convention, there was a dialogue between a stakeholder and citizens about drug treatments for neurodegenerative diseases. The problem was that the expert forgot to inform the citizens that he was linked to the pharmaceutical industry—the potential problems in situations such as these are clear.

A final limit, which may not even be a limit, is the surprising conception citizens have of bioethics, at least in this experience. As mentioned above, they both hope and fear emerging technology, and to solve this dilemma, they call on bioethics. They are strongly in favour of an ethical regulatory power, with normative power given to the ethical committees. That observation is very far from our French conception of ethics committees as strictly consultative, with the essential function not to decide but only to organise the social debate. It is further evidence of the heterogeneity of opinions that exists among international citizens.

4.   FOLLOW-UP

It remains too early to know what kind of influence such a citizen's debate will have on European policies. However, the European commission, which was represented by the head of the Science and Society Unit at the European conference, seems to be very concerned by the experience. However, within France, the national ethics committee linked with *Cité des Sciences et de l'Industrie* considers it crucially important to use this first experience to promote a democratic citizen debate if only at the national level. We are considering organising such a debate on one of the topics currently under discussion at the committee. It was quite surprising that the French national ethics committee was the only national ethics committee that took part in this experience since the support of an ethics committee could give a certain effectiveness to the citizen's deliberation.

In conclusion, this first experience of a European citizen's debate demonstrates the importance, in the future, of making citizens partners in order to achieve social acceptance of emerging technologies, especially about complex and controversial issues.

Interestingly, all the questions raised by *Meeting of the Minds*, which I have summarized here, are nothing less that the limits of direct democracy.

Developments in Genomics: Engaging Young People

Caroline HURREN

Public Engagement Development Group, Wellcome Trust, UK

The author would like to acknowledge the work of the grant holders whose projects are described in this paper.

The Wellcome Trust is an independent research-funding charity which was established in 1936 by Henry Wellcome and is funded from a private endowment. Its mission is to foster and promote research with the aim of improving human and animal health. This is mainly achieved by funding research in biomedical science but also in such diverse areas as biomedical ethics and the history of medicine.[1]

The Wellcome Trust has a major commitment to public engagement with science. One specific aspect of this work involves engaging young people. Today's young people will be the future citizens of this world. The applications arising from genomic research are likely to have a greater impact on the lives of these future citizens than on the lives of today's adults. It is therefore essential that young people are able to engage with the process and progress of biomedical science, including genomics.

Formal education, through schools, is an ideal environment in which to engage young people. However, schools are faced with a challenging dual agenda in science education. On the one hand, they are trying to equip young people with the interest, skills and knowledge to become scientists. And on the other, they are enabling all young people to develop the skills and understanding so that they can apply science to their personal lives now and in the future.

Within the UK, as well as in many other countries, this dual agenda has not been successfully delivered in the past. There are fewer young people choosing to study science beyond the compulsory age of 16,[2] and an increasing number of young people finding it difficult to see the relevance of science to their lives.[3] Research shows that young people find science more difficult than other subjects and many cite it as being boring.[4]

Our views as adults are influenced by what we learn at school. A survey by the market research company MORI in 2004[5] showed that 20% of the adult population was deterred from science because of the school approach to science, and this rises to 27% among people born after 1980. This is interesting as this timeframe correlates with the period when the school curriculum within England became much more prescriptive—which it remains today.

Three factors play an important part in how well schools can lay the foundations of science education:

- the continuing professional development that teachers undertake so that they are equipped to teach science well;
- the content of the curriculum and how it is assessed; and
- the availability of educational resources, support and other enrichment activities.

The Wellcome Trust is active in all these areas, supporting a range of initiatives (Figure 1). An example of three projects is given below, each of which addresses one of these factors.

1.   CONTINUING PROFESSIONAL DEVELOPMENT

Unlike other school subjects, science moves on at a rapid pace, so teachers need to continually keep up-to-date with these developments through continuing professional development.

Recent research commissioned by the Wellcome Trust[6] shows that most science teachers recognise this need to keep up-to-date. However, many have had no professional training relating to their subject in the last five years as this is not part of the school culture. Training for teachers more usually focuses on issues such as managing budgets or managing poor behaviour in the classroom—not on the subject that is being taught.

The Wellcome Trust is partnering with the British Government in funding a major initiative to create a network of professional development centres for school science educators.

The national network of Science Learning Centres[7] offers high-quality professional development for those involved in science education, including secondary school science teachers, technicians and primary school teachers.

The network enables those working in science education to access cutting-edge technology and leading scientific research. The aim is to support teachers in delivering intellectually stimulating and relevant science education and to help them stay in touch with developments in science.

The Science Learning Centres network operates through nine regional Centres, funded by the Government and the Wellcome-funded National Centre.

The inclusion of contemporary science and of issues raised, including genomics, has been addressed by a number of the courses both at the regional and national centres. One example is given in Figure 2.

2. CURRICULUM

The Wellcome Trust has played a significant part, along with the Nuffield Foundation and Salters Institute, in supporting the development of a new way of teaching science from age 14 to 16, called 21st Century Science.[8] All English schools will teach a version of this curriculum from September 2006 which attempts to address the dual agenda of science education. Initial evaluation from the pilot project has been positive.[9]

The approach offers all students the chance to study a course which develops scientific literacy, with two main strands:

- key science explanations that help make sense of our lives, and
- ideas about science that show how science works.

This course views science from the perspective of a member of the public and is taught in the context of topics of current and cultural interest, including: you and your genes, air quality, earth in the universe, keeping healthy, material choices, radiation and life, life on earth, food matters, radioactive materials.

This course, for all students, is augmented by one which covers fundamental principles about science for those who have more of a bent towards science and might want to train as a scientist in the future. There is also a course which trains people for science-related careers such as technical jobs.

3.   RESOURCES

The third aspect of science education is to ensure that the classroom experience is enriched by the availability of appropriate educational resources and support. These resources may either be aimed at young people themselves or at teachers. One example for which the Wellcome Trust has a significant programme of funding is the use of theatre, film and performance to engage young people.

One project funded by the Wellcome Trust, called IMPACT Danscience, made use of the unlikely approach of using dance to engage students with epigenetics.[10] The project united a British Indian contemporary dance company, a biomedical scientist and eight young dancers aged 13 to 18 from South London.

The dance combines a popular traditional dance style from South India, called *Bharata Natyam,* with the individual dance styles of each of the participants, including elements of street dance and hip hop as well as more contemporary styles such as ballet and jazz.

In traditional Bharata Natyam performances, each movement has a distinct meaning and each dance tells a story. In this dance, each movement has a meaning signifying the process of cell division, mitosis and the DNA double helix with the dancers representing chromosomes. There is a recurring theme or "motif" in the dancers' movements. This "motif" is then developed in different ways as a variation on the theme. Figure 3 sets this out in more detail.

4.   CONCLUSION

Public engagement with science is a much discussed topic among researchers, policymakers, educators and science communicators. Achieving genuine engagement, however, is a complicated proposition, one that requires long-term commitment and broad-based support. The Wellcome Trust recognizes the inevitable benefits of this effort; increased public participation in scientific discussion will improve both the quality and usefulness of new developments such as genomics. By focusing on young people, the Trust is attempting to achieve the most long-lasting impact possible for the future. The three initiatives described in this chapter have provided a snapshot of the Trust's significant effort to increase young people's engagement through formal education in schools. Through projects like these, the Wellcome Trust and its grant holders will contribute to public engagement as they affirm their commitment to science and to society.

FIGURE 1: FORMAL SCIENCE EDUCATION

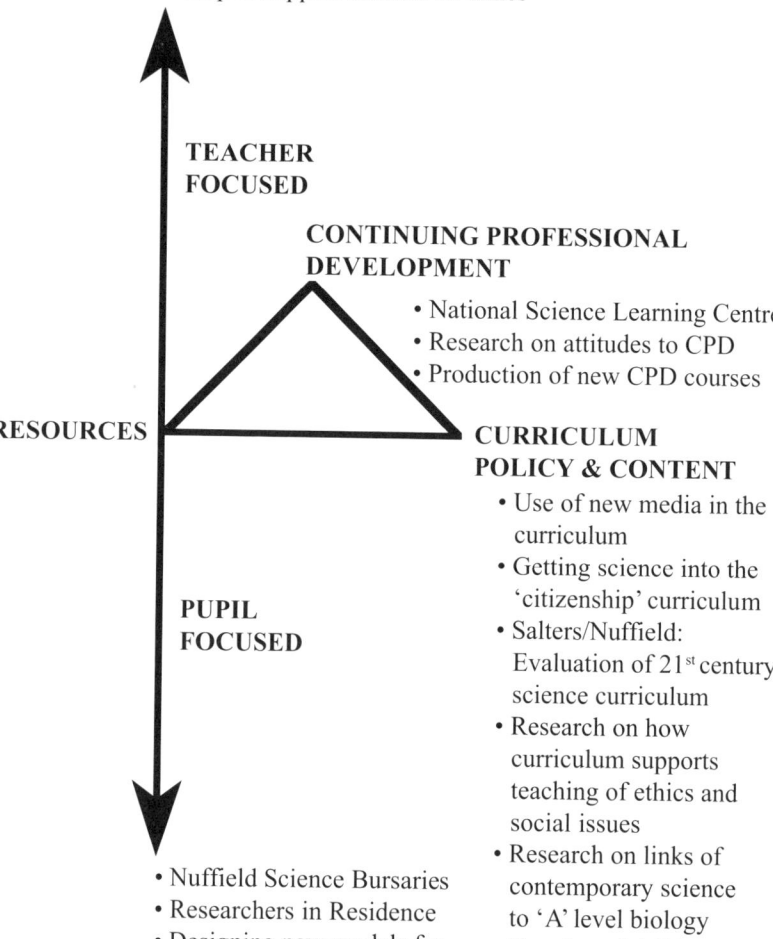

- Using fictional TV to teach science
- Resources on contemporary topics
- Help to support debates on ethics

TEACHER FOCUSED

CONTINUING PROFESSIONAL DEVELOPMENT

- National Science Learning Centre
- Research on attitudes to CPD
- Production of new CPD courses

RESOURCES

CURRICULUM POLICY & CONTENT

- Use of new media in the curriculum
- Getting science into the 'citizenship' curriculum
- Salters/Nuffield: Evaluation of 21st century science curriculum
- Research on how curriculum supports teaching of ethics and social issues
- Research on links of contemporary science to 'A' level biology
- Creation of A/S Level in history + philosophy of science

PUPIL FOCUSED

- Nuffield Science Bursaries
- Researchers in Residence
- Designing new models for debate
- Theatre, film & performance in schools
- National debating competitions
- Junior café scientifique

FIGURE 2: HOW SCIENCE WORKS: BRINGING CONTEMPORARY AND CONTROVERSIAL SCIENCE INTO THE CLASSROOM

"Teachers should ensure that the knowledge, skills and understanding of how science works are integrated into the teaching of the breadth of study."

National Curriculum Programme of Study: Science Key stage 4

This course will provide teachers with opportunities to explore how pupils can be taught:

- About the use of contemporary science and its benefits, drawbacks and risks

- How decisions are made about science including socio-ethical issues

- About how scientific ideas change over time and the role of the scientific community in validating these changes.

Teachers need time to engage with these new concepts from the programme of study, especially to consider where they will find information, ideas and resources to support their scheme of work. Teachers also need to enhance their expertise with implementing these strategies in the classroom.

The course offers teachers of all science disciplines a rich menu of activities.

The common areas of the course comprise:

- How science research is perceived by the public

- Developing more effective approaches to teaching about scientific controversy

- Time to plan school based activities

- Time to explore new teaching and learning resources

- Time for teachers of different disciplines to share with each other from the optional strands of the course.

Teachers are invited to choose one of the following strands:

- Medical imaging (physics)

- Climate change (chemistry)

- Beyond the genome (biology)

The strands will involve contact with research scientists at the forefront of these topical fields, visits to research centres and practical work. There will also be sessions in which leading teaching specialists will steer participants to consider the implications of the contemporary science for classroom practice.

MEDICAL IMAGING

This strand is expected to appeal mainly to physics specialists.

In addition to following the common areas of the course, participants will explore the range of modalities for modern medical imaging, through presentations from scientists and technicians in the field. They will consider what can be achieved to demonstrate the uses of imaging techniques and the physics behind them through classroom demonstrations, practicals and problem-based learning. The ethical and moral dilemmas associated with the tension between medical practice and health economics, and with the use of humans as experimental subjects will be debated. In addition, there will be a site visit to a world-class research facility specialising in neuro-imaging, to look at the physics behind these techniques and the possibilities they raise for cutting-edge research.

CLIMATE CHANGE

This strand is expected to appeal mainly to chemistry specialists.

In addition to following the common areas of the course, participants will explore the "Big Ideas" on climate change, through presentations from scientists in the field. These will include monitoring air quality, controlling carbon emissions and molecular models of the greenhouse effect. Participants will analyse the conceptual challenges of the carbon cycle, and consider practical ways to demonstrate the underpinning science in the classroom. The dilemmas surrounding our options for controlling climate change will be debated, modelling classroom practice, and they will consider climate change as a context for exploring with students the relationship between correlation and causation.

BEYOND THE GENOME

This strand is expected to appeal mainly to biology specialists.

In addition to following the common areas of the course, participants will explore cutting-edge aspects of post-genomic biology, through presentations from scientists in the field. These will include the future for crop research in a GM-averse society; genome knowledge and our understanding of microbial disease; and (with input from the Sanger Institute) the shift from gene therapy to the new hope of pharmacogenetics for human health. In addition, there will be a site visit to follow the story of a genetic research proposal from grant application through experiments to publication. Participants will examine practical work which introduces pupils to gene technology in ways that reinforce basic knowledge. They will also experience strategies for exploring the ethical dilemmas associated with the exploitation of genetic knowledge.

FIGURE 3: IMPACT DANSCIENCE

CYCLE 1

The cell: The dancers represent *chromosomes* inside a cell's *nucleus*. It looks as if they are moving randomly but patterns form.

Mitosis: The dancers perform in pairs to show how chromosomes divide & split into two identical *daughter* cells.

DNA: Part of the material inside the chromosomes is magnified to show DNA. The dancers do a "DNA walk" to demonstrate the shape of the double helix.

CYCLE 2

The structure of Cycle 1 is repeated (i.e. cell, mitosis, DNA), but this time the cell and its DNA are gradually being affected by the environment

CYCLE 3

The structure repeats again, but here the 'abnormality' caused by the previous cycle's effects on the DNA are shown—the epigenetic effect.

The abnormalities no longer occur because of direct environmental influences but because they have been passed down from the previous cycle.

CYCLE 4

The dance is open-ended. Will the epigenetic effects be passed on to the next cell or perhaps the next generation?

MUSIC

A Tibetan bowl introduces the first cycle and the music conjures up a happy, positive mood to correspond with the healthy DNA.

In the second cycle the music gradually becomes darker as we hear external sounds, such as mobile phones, aeroplanes, traffic jams. The atmosphere of the music is designed to give a feeling of danger for the DNA and the process of cell division. As well as conveying the way in which our environment affects our genes.

The third cycle, marked by eerie and sometimes sorrowful music, portrays the disturbance and chaos within the DNA, brought on by the effects of the external environment.

The piece finishes with an uncertain quality which matches the open ending of the dance.

REFERENCES

1. Wellcome Trust, Strategic Plan 2005–2010: Making a Difference (2005).
2. Department for Education and Skills, Statistics of Education: Education and Training Statistics for the UK (various years).
3. J. Osborne & S. Collins, Pupils' & Parents' Views of the School Science Curriculum (2000).
4. OCR, Students and Science Report (2006).
5. Office of Science and Technology, Department of Trade and Industry, Science in Society: Findings from Qualitative and Quantitative Research (2005).
6. The Wellcome Trust, Believers, Seekers and Sceptics: What Teachers Think About Continuing Professional Development (2006).
7. Science Learning Centres, http://www.sciencelearningcentres.co.uk.
8. 21st Century Science, http://www.21stcenturyscience.org; J. Burden, *Twenty First Century Science*, 213 EDUCATION IN SCIENCE, 10 (2005).
9. R. Millar, *Science in Education: Implications for Formal Education*, in Engaging Science: Thoughts, Deeds, Analysis and Action (J. Turney ed., 2006).
10. IMPACT Danscience, http://www.impactdanscience.co.uk.

Interdisciplinary Graduate Training in Public Health Genetics at the University of Washington

Melissa A. AUSTIN

School of Public Health and Community Medicine
University of Washington, Seattle, Washington, USA

1. INSTITUTE FOR PUBLIC HEALTH GENETICS

Public health genetics, the application of the rapid advances in human genetics and genomic science to improve public health and prevent disease, is an emerging field on a global scale. As a result, there is an urgent need to educate public health academicians and practitioners in this highly interdisciplinary field. To meet this need, the University of Washington established the Institute for Public Health Genetics (IPHG) in 1997, and now has permanent state funding. The mission of the IPHG is to provide broad, interdisciplinary training for future public health genetics professionals, to facilitate research in this emerging field, and to serve as a resource for continuing professional education. The Institute is a collaborative effort that involves faculty members and administrators from seven different schools and colleges at the University of Washington, and includes active relationships with the Washington State Department of Health and the Fred Hutchinson Cancer Research Center. The IPHG's unique academic programme integrates genomics with the public health science disciplines of epidemiology, biostatistics, environmental health, and health services research, and with bioethics, social sciences, law, public policy and health economics, and provides graduate training at both the masters and doctoral levels.

2.   FUNDAMENTAL AREAS OF STUDY AND CORE KNOWLEDGE AREAS

To provide a framework for this interdisciplinary training programme, the IPHG curriculum is divided into Fundamental Areas of Study and two broad Core Knowledge Areas. The Fundamental Areas of Study include human genetics, genomics, and public health. The Core Knowledge Areas are: (A) Genomics in Public Health, including genetic epidemiology, ecogenetics (gene-environment interactions) and pharmacogenetics; and (B) Implications of Genomics for Society, including ethics, social science, law, public policy, health economics, and outcomes research. Fifteen graduate courses, many team-taught by core faculty members, cover all of these areas and form the basis of the degree programmes. A bi-weekly interactive seminar brings together students from all of the degree programmes and the core faculty, and features a wide variety of speakers from numerous disciplines related to public health genetics.

3.   GRADUATE DEGREE PROGRAMMES

As described briefly below, the University of Washington currently offers three interdisciplinary graduate programmes in Public Health Genetics: an MPH, a Ph.D., and a Graduate Certificate.

3.1   *Master of Public Health (MPH) in Public Health Genetics*

The University of Washington offers the only accredited MPH in Public Health Genetics in the United States. The curriculum includes two years of coursework in the Fundamental Areas of Study and Core Knowledge Areas described above, a practicum experience, and a research master's thesis.

As of this writing, 57 MPH students have entered the programme and 36 have graduated. The majority of students have undergraduate degrees in biology, biochemistry, genetics, or molecular biology, but others have backgrounds ranging from nursing and medicine to law and philosophy.

Approximately half of the MPH graduates have undertaken more advanced degree training, including programmes in genetic counselling, law school, medical school, and Ph.D. programmes, including seven in the University of Washington Ph.D. programme in Public Health Genetics. Several students have taken fellowship opportunities at the Centers for Disease Control and Prevention in Atlanta, and others have positions in state health departments, at the NIH, and with the Secretary's Advisory Committee on Genetic Testing.

3.2   *Ph.D. in Public Health Genetics*

The Ph.D. programme in Public Health Genetics was initiated during the 2003-2004 academic year. The overall goals of this programme are:

1. To train researchers, educators, and programme administrators for careers in academic institutions, healthcare delivery systems, public health departments, government agencies and the private sector.
2. To provide interdisciplinary education so that graduates can address scientific and policy questions from a variety of perspectives.

At this time, there are a total of 16 students pursuing a Ph.D. in Public Health Genetics, and we anticipate the first graduates during the 2006-2007 academic year.

A unique component of the Ph.D. programme is the preliminary examination, designed to be taken after students have completed the required core courses, usually at the end of the second year of study. The purpose of the examination is for students to demonstrate competency in each of the Core Knowledge Areas, before initiating their dissertation project. The examination is written as a collaborative project with participation from all core faculty members. It is intended to be comprehensive and integrative, and uses a case study approach with questions relating to each component of the Core Knowledge Areas. Upon passing this examination, students develop and complete their dissertation projects, undertake the general examination, and defend their dissertations. The dissertation project is required to have components of both Core Knowledge Areas. Examples of dissertation projects currently being developed are Epidermal Growth Factor Receptor Pharmacogenomics: an Economic and Policy Evaluation; and Genetic and Behavioral Risk Factors for Pancreatic Cancer: Association of Diabetes Candidate Genes with Pancreatic Cancer Risk and Policy Analysis of "Reduced Harm" Tobacco Products.

3.3   *Graduate Certificate in Public Health Genetics*

The Graduate Certificate Programme is designed for students currently enrolled in any other graduate degree programme at the University of Washington who wish to learn about public health genetics. Certificate students are required to take three IPHG core courses and the interactive seminar. Upon graduation, they receive a paper certificate and an acknowledgment of this training on their official transcript. Currently, a total of 30 students from 10 different departments have been accepted into the certificate programme, and 25 have completed the requirements to date.

4.   INTERDISCIPLINARY FACULTY

The IPHG involves 16 core faculty members from seven different schools and colleges at the University of Washington: the School of Public Health and Community Medicine (SPHCM), the School of Law, the School of Medicine, the College of Arts and Sciences, the School of Pharmacy, the School of Nursing, and the Daniel J. Evans School of Public Affairs. In addition, active collaborative relationships continue with the Washington State Department of Health and the Fred Hutchinson Cancer Research Center.

All IPHG core faculty members serve on the Academic Programme Committee that oversees the academic aspects of the programme. Members of this committee teach IPHG courses, coordinate curriculum matters, decide student admissions for the degree programmes, provide student advising and mentoring, review MPH thesis topics, write and grade the Ph.D. preliminary exam, and serve on masters thesis committees and doctoral dissertation supervisory committees. However, since all core faculty members have their primary academic appointments in other departments and schools, most undertake these responsibilities in addition to their usual departmental expectations, reflecting their unusual commitment to the programme.

In addition to core faculty members, we recognize a continuing need to identify researchers and health professionals who are interested in public health genetics, and provide them with an affiliation to the IPHG. Thus, we have designated a category of IPHG faculty "members" who may become involved in a variety of activities, including mentoring students, serving on thesis committees, providing practicum sites, giving an occasional guest lecture or seminar, and/or participating in outreach activities and conferences. At present there are 33 such IPHG faculty members from a variety of departments at the University of Washington, from Children's Hospital and Medical Center, from the Washington State Department of Health, and from the Fred Hutchinson Cancer Research Center. To complete the governance of the IPHG, there is also an Internal Advisory Board that consists of deans and chairs of the schools and departments involved with the IPHG.

5.   CHALLENGES TO INTERDISCIPLINARY TRAINING PROGRAMMES

The IPHG is one of many interdisciplinary research and training programmes at the University of Washington. Although these programmes represent some of the most innovative programmes within the University, the current administrative structure, built on traditional schools and departments, presents several challenges for their success. To address these challenges, the Graduate School at the

University of Washington established the Network of Interdisciplinary Initiatives (NII). The NII meets quarterly and includes numerous programmes from across campus. It has identified three major areas in which policy changes could significantly increase support for interdisciplinary teaching, research, and training. These areas are: 1) faculty appointments, promotion, and tenure; 2) allocation of resources, including indirect costs; and 3) development fundraising, and outreach. The IPHG participates actively in this network, working towards new University policies that will encourage new interdisciplinary programmes, as well providing ongoing support to sustain established programmes.

6.   RESEARCH PROGRAMMES

Despite these challenges, the IPHG core faculty have been highly successful in leveraging their involvement with the Institute to obtain grant funding from a variety of sources. These grants include the Northwest Center for Genomics and Public Health funded by the Centers for Disease Control and Prevention (CDC), the UW Center of Excellence in Ethical, Legal and Social Implications Research funded by the National Human Genome Research Institute (NHGRI), the Ethical Legal and Social Implications Core of the Center for Ecogenetics and Environmental Health funded by the National Institute of Environmental Health Sciences (NIEHS), and several grants from the Health Resources and Services Administration (HRSA) grants.. All of these grants are integrated with the graduate training programme by providing unique, interdisciplinary research opportunities for graduate students, who in turn provide support for the faculty research agendas.

7.   CONCLUSION

A major challenge to public health is to develop policies and procedures that maximize the population health benefits from advances in genomics, while ensuring that genetic information is not misused. Thus, graduate training in Public Health Genetics must encompasses a broad range of disciplines and apply a truly interdisciplinary approach, including participation from scientists, ethicists, social sciences, economists, and legal and policy experts. Unlike most current specialists in these fields who were primarily trained in a single discipline, this new generation of public health professionals will need to integrate several traditional disciplines into their work, and will need to appreciate the value of diverse perspectives in solving these complex problems. Finally, education in public health genetics must be ongoing, so that the potential of promising genetic technologies can be used to benefit the health of all.

ACKNOWLEDGMENTS

This work was supported by University Initiative Fund of the University of Washington and NIH grant P30ES07033 from NIEHS.